GLENCOE MATHEMATICS

Pre-Algebra

Parent and Student Study Guide Workbook

Glencoe McGraw-Hill

New York, New York
Columbus, Ohio
Chicago, Illinois
Peoria, Illinois
Woodland Hills, California

Glencoe/McGraw-Hill

A Division of The McGraw-Hill Companies

Copyright © by the McGraw-Hill Companies, Inc. All rights reserved.
Printed in the United States of America. Permission is granted to reproduce the material contained herein on the condition that such material be reproduced only for classroom use; be provided to students, teachers, and families without charge; and be used solely in conjunction with Glencoe Pre-Algebra. Any other reproduction, for use or sale, is prohibited without prior written permission of the publisher.

Send all inquiries to:
Glencoe/McGraw-Hill
8787 Orion Place
Columbus, OH 43240

ISBN: 0-07-827786-8

Pre-Algebra
Parent and Student Study Guide Workbook

5 6 7 8 9 10 079 11 10 09 08 07 06

Contents

Chapter	Title	Page
	To the Parents of *Glencoe Pre-Algebra Students* iv	
1	The Tools of Algebra 1	
2	Integers 9	
3	Equations 16	
4	Factors and Fractions 24	
5	Rational Numbers 33	
6	Ratio, Proportion, and Percent 44	
7	Equations and Inequalities 54	
8	Functions and Graphing 61	
9	Real Numbers and Right Triangles 72	
10	Two-Dimensional Figures 81	
11	Three-Dimensional Figures 90	
12	More Statistics and Probability 98	
13	Polynomials and Nonlinear Functions 108	
	Answer Key for Chapter Reviews 115	

To the Parents of *Glencoe Pre-Algebra* Students

You teach your children all the time. You taught language to your infants and you read to your son or daughter. You taught them how to count and use basic arithmetic. Here are some ways you can continue to reinforce mathematics learning.

- Encourage a positive attitude toward mathematics.
- Set aside a place and a time for homework.
- Be sure your child understands the importance of mathematics achievement.

The *Glencoe Pre-Algebra* **Parent and Student Study Guide Workbook** is designed to help you support, monitor, and improve your child's math performance. These worksheets are written so that you do not have to be a mathematician to help your child.

The **Parent and Student Study Guide Workbook** includes:

- A 1-page *worksheet* for every lesson in the Student Edition (101 in all). Completing a worksheet with your child will reinforce the concepts and skills your child is learning in math class. Upside-down answers are provided right on the page.
- A 1-page *chapter review* (13 in all) for each chapter. These worksheets review the skills and concepts needed for success on tests and quizzes. Answers are located on pages 115–119.

Online Resources
For your convenience, these worksheets are also available in a printable format at **www.pre-alg.com/parent_student**.

Pre-Algebra Online Study Tools can help your student succeed.

- **www.pre-alg.com/extra_examples**
  shows you additional worked-out examples that mimic the ones in the textbook.
- **www.pre-alg.com/self_check_quiz**
  provides a self-checking practice quiz for each lesson.
- **www.pre-alg.com/vocabulary_review**
  checks your understanding of the terms and definitions used in each chapter.
- **www.pre-alg.com/chapter_test**
  allows you to take a self-checking test before the actual test.
- **www.pre-alg.com/standardized_test**
  is another way to brush up on your standardized test-taking skills.

1-1 Using a Problem-Solving Plan (Pages 6–10)

You can use a four-step plan to solve real-life, math-related problems.

Explore	Read the problem carefully. Ask yourself questions like "What facts do I know?" and "What do I need to find out?"
Plan	See how the facts relate to each other. Make a plan for solving the problem. Estimate the answer.
Solve	Use your plan to solve the problem. If your plan does not work, revise it or make a new one.
Examine	Reread the problem. Ask, "Is my answer close to my estimate? Does my answer make sense for the problem?" If not, solve the problem another way.

Example

Luther bought 8 CDs at a sale. The first CD purchased costs $13, and each additional CD costs $6. What was the total cost before tax?

Explore	You are given the cost of the first CD and the cost of additional CDs. You need to find the total cost.
Plan	First find the number of additional CDs after the first CD he purchased. Multiply that number by $6 and add $13 for the first CD. Estimate the total cost by using $15 + 7 × $5 = $50.
Solve	8 − 1 = 7, 7 × $6 = $42, $42 + $13 = $55
Examine	The total cost of $55 is close to the estimate of $50, so the answer is reasonable.

Practice

1. The table at the right shows estimates of the number of species of plants and animals on Earth. Find the total number of species on Earth.

Group	Number
Mammals, Reptiles, Amphibians	13,644
Birds	9,000
Fish	22,000
Plants	443,644
Invertebrates	4,400,000

    a. Write the explore step.
    b. Write the plan step.
    c. Solve the problem.
    d. Examine your solution. Is it reasonable?

2. Jeff is 10 years old. His younger brother, Ben, is 4 years old. How old will Jeff be when he is twice as old as Ben?

3. **Standardized Test Practice** At Camp Mystic, there are 576 campers. If 320 campers are boys, then how many campers are girls?

    A 432 girls B 320 girls C 256 girls D 144 girls

Answers: 1. See Answer Key. 2. 12 years old 3. C

© Glencoe/McGraw-Hill   1   Glencoe Pre-Algebra

NAME _____ DATE _____ PERIOD _____

1-2 Numbers and Expressions *(Pages 12–16)*

A mathematical **expression** is any combination of numbers and operations such as addition, subtraction, multiplication, and division. To **evaluate** an expression, you find its numerical value. To avoid confusion, mathematicians established the order of operations to tell us how to find the value of an expression that contains more than one operation.

Order of Operations	1. Do all operations within grouping symbols first; start with the innermost grouping symbols. Grouping symbols include **parentheses**, ( ), and **brackets**, [ ].
	2. Next, do all multiplications and divisions from left to right.
	3. Then, do all additions and subtractions from left to right.

Examples Find the value of each expression.

a. $7 + 8 \div 2 - 5$

$7 + 8 \div 2 - 5$
$= 7 + 4 - 5$ Do multiplications and divisions first.
$= 11 - 5$ Add and subtract from left to right.
$= 6$

b. $3[(4 + 5) \div (15 - 12)] + 8$

$3[(4 + 5) \div (15 - 12)] + 8$
$= 3[9 \div 3] + 8$ Do operations in innermost grouping symbols first.
$= 3[3] + 8$ Multiply, then add.
$= 17$

Try These Together

Find the value of each expression.

1. $17 + 4 \cdot 8$
2. $16 \div 4 + 24 \div 8$
3. $3 + 8(2 + 4)$

HINT: Remember to follow the order of operations when finding each value.

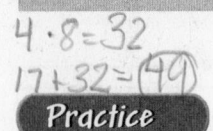

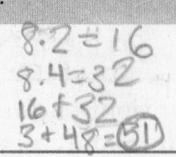

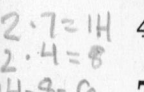

Practice

Find the value of each expression.

4. $2(7 - 4) \div 6$
5. $14 - (9 \div 3)$
6. $5 \cdot 6 - 12$
7. $[3(14 \div 7) + 2 \cdot 8] \div 11$
8. $2(3 \cdot 4) \div 6 - 2\left(\dfrac{6}{3}\right) \div 2$
9. $\dfrac{9 + 6}{30 - 27}$
10. $18 + (16 - 9) \cdot 4$
11. $42 - 7 \cdot 4$
12. $2[7(3 - 2) + 4(10 - 8)]$
13. $11[2(18 - 13) - 4 \cdot 2]$
14. $7[10(17 - 2) - 8(6 \div 2)]$
15. $4[3(10 - 7) + (11 \cdot 2)]$

16. **Standardized Test Practice** At a garage sale, Doug earns $2 for each book he sells, and Linda earns $3 for each used CD that she sells. Doug sells 15 books and Linda sells 12 CDs. They share the total earnings equally. What is each person's share of the earnings?

A $66   B $36   C $33   D $30

Answers: 1. 49   2. 7   3. 51   4. 1   5. 11   6. 18   7. 2   8. 2   9. 5   10. 46   11. 14   12. 30   13. 22   14. 882   15. 124   16. C

1-3 Variables and Expressions (Pages 17–21)

Aside from the operation symbols you already know, algebra uses placeholders, usually letters, called **variables**. The letter x is used very often as a variable in algebra, but variables can be any letter. An expression such as $a \div 2 + 110$ is an **algebraic expression** because it is a combination of variables, numbers, and at least one operation. You can evaluate algebraic expressions by replacing the variables with numbers and then finding the numerical value of the expression.

Substitution Property of Equality	For all numbers a and b, if $a = b$, then a may be replaced by b.
Special Notation	$3d$ means $3 \times d$ 　　　　$7st$ means $7 \times s \times t$ xy means $x \times y$ 　　　　$\frac{q}{4}$ means $q \div 4$

Examples Find the value of each expression.

a. Evaluate $a + 47$ if $a = 12$.
$a + 47 = 12 + 47$ Replace a with 12.
$= 59$

b. Evaluate $\frac{7r}{2}$ if $r = 4$.
$\frac{7r}{2} = \frac{7(4)}{2}$ Replace r with 4.
$= \frac{28}{2}$ or 14

Practice

Evaluate each expression if $x = 2$, $y = 7$, and $z = 4$.

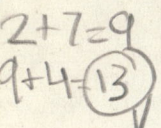

1. $x + y + z$
2. $(z - x) + y$
3. $2x - z$
4. $4y - 3z$
5. $4(x + y) \div z$
6. $4x + 2y$
7. $8 + 10 \div x + z$
8. $y + 2z \div 3$
9. $\frac{2x + 2y}{6}$

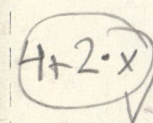

Translate each phrase into an algebraic expression.

10. 4 more than 2 times a number
11. the product of x and y
12. the quotient of 16 and a
13. the sum of m and 8 divided by 2

14. **Standardized Test Practice** The carrying capacity of an environment is the number of individuals the natural ecosystem of an area is able to support. If one mouse requires 1.6 acres of land for survival, what is the carrying capacity of a 528-acre park for mice?
  A 845 mice　　**B** 528 mice　　**C** 330 mice　　**D** 33 mice

Answers: 1. 13 2. 9 3. 0 4. 16 5. 9 6. 22 7. 17 8. 9$\frac{2}{3}$ 9. 3 10. 4 + 2x 11. xy 12. $\frac{16}{a}$ 13. $\frac{m+8}{2}$ 14. C

1-4 Properties (Pages 23–27)

Commutative Property of Addition and Multiplication	
The order in which numbers are added does not change the sum. $5 + 3 = 3 + 5$ For any numbers a and b, $a + b = b + a$.	The order in which numbers are multiplied does not change the product. $2 \cdot 4 = 4 \cdot 2$ For any numbers a and b, $a \cdot b = b \cdot a$.
Associative Property of Addition and Multiplication	
The way in which addends are grouped does not change the sum. $(2 + 3) + 4 = 2 + (3 + 4)$ For any numbers a, b, and c, $(a + b) + c = a + (b + c)$.	The way in which factors are grouped does not change the product. $(2 \cdot 3) \cdot 4 = 2 \cdot (3 \cdot 4)$ For any numbers a, b, and c, $(a \cdot b) \cdot c = a \cdot (b \cdot c)$.
Identity Property of Addition and Multiplication	
The sum of an addend and zero is the addend. $6 + 0 = 6$ For any number a, $a + 0 = a$.	The product of a factor and one is the factor. $6 \cdot 1 = 6$ For any number a, $a \cdot 1 = a$.
Multiplicative Property of Zero	
The product of a factor and zero is zero. $5 \cdot 0 = 0$. For any number a, $a \cdot 0 = 0$.	

Practice

Name the property shown by each statement.

1. $x \cdot 0 = 0$
2. $a + 8 = 8 + a$
3. $2x(y) = 2xy$
4. $m + 0 = m$
5. $3(x + y) = (x + y)3$
6. $(4c)d = 4(cd)$
7. $7x + 10 = 10 + 7x$
8. $4x \cdot 1 = 4x$
9. $10x + 8y = 8y + 10x$

Find each sum or product mentally using the properties above.

10. $37 + 8 + 23$
11. $5 \cdot 11 \cdot 2$
12. $4 \cdot 12 \cdot 6 \cdot 0$

13. Rewrite $18y \cdot 4x$ using the Commutative Property.

14. Rewrite $(2x + 8) + 4$ using the Associative Property. Then simplify.

15. **Standardized Test Practice** Juana is 4 feet 8 inches tall. She won 1st place in a cross-country race. To receive her medal, she stood on a platform that was 18 inches tall. What was the total distance from the top of Juana's head to the ground when she was standing on the platform?
   - **A** 5 feet 6 inches
   - **B** 5 feet 8 inches
   - **C** 6 feet
   - **D** 6 feet 2 inches

Answers: 1. multiplicative property of zero 2. commutative 3. associative 4. identity 5. commutative 6. associative 7. commutative 8. identity 9. commutative 10. 68 11. 110 12. 0 13. $4x \cdot 18y$ 14. $2x + (8 + 4)$; $2x + 12$ 15. D

NAME _____ DATE _____ PERIOD _____

1-5 Variables and Equations (Pages 28–32)

A mathematical sentence such as $2001 - 1492 = 509$ is called an **equation**. An equation that contains a variable is called an **open sentence**. When the variable in an open sentence is replaced with a number, the sentence may be true or false. A value for the variable that makes an equation true is called a **solution** of the equation. The process of finding a solution is called **solving the equation**.

Examples Identify the solution to each equation from the list given.

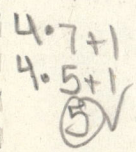

a. $13 + s = 72$; 48, 53, 59

Replace s with each of the possible solutions to solve the equation.

$13 + 48 = 72$
 $61 = 72$ False. 48 is not a solution.
$13 + 53 = 72$
 $66 = 72$ False. 53 is not a solution.
$13 + 59 = 72$
 $72 = 72$ True. 59 is the solution to the equation.

b. $3y - 2 = 4$; 1, 2

Replace y with each of the possible solutions to solve the equation.

$3(1) - 2 = 4$, or $3 - 2 = 4$
 $1 = 4$ False. 1 is not the solution.
$3(2) - 2 = 4$, or $6 - 2 = 4$
 $4 = 4$ True. 2 is the solution.

Try These Together
Identify the solution to each equation from the list given.

1. $15 - 8 = x$; 23, 10, 7
2. $6 = \frac{24}{p}$; 8, 6, 4

HINT: Replace the variable with each possible solution to see if it makes the open sentence true.

Practice
Identify the solution to each equation from the list given.

3. $4x + 1 = 21$; 7, 5, 4
4. $98 - c = 74$; 24, 30, 34
5. $7 = \frac{x}{4}$; 28, 30, 32
6. $82 + a = 114$; 62, 32, 22
7. $19 = a + 7$; 17, 12, 8
8. $6x = 48$; 6, 7, 8

Solve each equation mentally.

9. $n + 6 = 12$
10. $56 = 7j$
11. $y - 17 = 41$
12. $\frac{32}{k} = 4$
13. $10 + p = 17$
14. $6m = 48$

15. **Standardized Test Practice** Sanford and Audrey are driving 65 miles per hour. If they travel 358 miles without stopping or slowing down, about how long will their trip take?

A 4.5 hours B 5.0 hours C 5.5 hours D 6.0 hours

Answers: 1. 7 2. 4 3. 5 4. 24 5. 28 6. 32 7. 12 8. 8 9. 6 10. 8 11. 58 12. 8 13. 7 14. 8 15. C

© Glencoe/McGraw-Hill   Glencoe Pre-Algebra

NAME _____ DATE _____ PERIOD ____

1-6 Ordered Pairs and Relations *(Pages 33–38)*

In mathematics, you can locate a point by using a **coordinate system**. The coordinate system is formed by the intersection of two number lines that meet at their zero points. This point is called the **origin**. The horizontal number line is called the ***x*-axis** and the vertical number line is called the ***y*-axis**.

You can graph any point on a coordinate system by using an **ordered pair** of numbers. The first number in the pair is called the ***x*-coordinate** and the second number is called the ***y*-coordinate**. The coordinates are your directions to the point.

Example

Graph the ordered pair (4, 3).
- Begin at the origin. The x-coordinate is 4. This tells you to go 4 units right of the origin.
- The y-coordinate is 3. This tells you to go up three units.
- Draw a dot. You have now graphed the point whose coordinates are (4, 3).

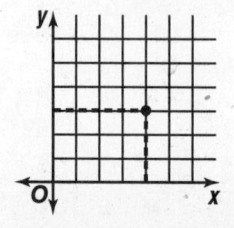

Try These Together

Use the grid below to name the point for each ordered pair.

1. (2, 1) **2.** (0, 2)

Hint: The first number is the x-coordinate and the second number is the y-coordinate.

Practice

Use the grid at the right to name the point for each ordered pair.

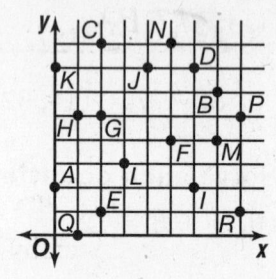

3. (5, 4)   **4.** (6, 7)   **5.** (7, 6)
6. (2, 5)   **7.** (1, 5)   **8.** (6, 2)

Use the grid to name the ordered pair for each point.

9. K   **10.** C   **11.** Q   **12.** L
13. N   **14.** P   **15.** J   **16.** M

17. **Standardized Test Practice** On the grid above, what would you have to do to the ordered pair for point *R* to get the ordered pair for point *P*?
 A Add 4 to the *x*-coordinate.
 B Add 4 to the *y*-coordinate.
 C Subtract 4 from the *x*-coordinate.
 D Subtract 4 from the *y*-coordinate.

Answers: 1. E 2. A 3. F 4. D 5. B 6. G 7. H 8. I 9. (0, 7) 10. (2, 8) 11. (1, 0) 12. (3, 3) 13. (5, 8) 14. (8, 5) 15. (4, 7) 16. (7, 4) 17. B

© Glencoe/McGraw-Hill   Glencoe Pre-Algebra

NAME _____ DATE _____ PERIOD ___

1-7 Scatter Plots (Pages 40–44)

A **scatter plot** is a graph consisting of isolated points that shows the general relationship between two sets of data.

Finding a Relationship for a Scatter Plot	**positive relationship:** points suggest a line slanting upward to the right
	negative relationship: points suggest a line slanting downward to the right
	no relationship: points seem to be random

Example

What type of relationship does this graph show?

Notice that the points seem to suggest a line that slants upward to the right, so this graph shows a positive relationship between a person's age and their height.

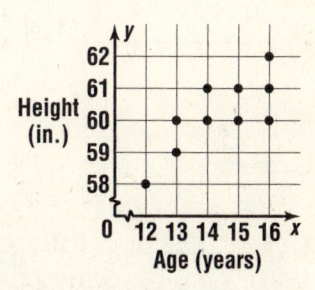

Try These Together

What type of relationship, positive, negative, or none, is shown by each scatter plot?

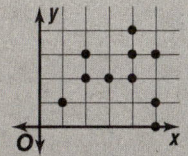

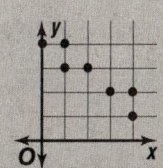

1.
2.

HINT: Notice that the points in Exercise 1 seem scattered while those in Exercise 2 suggest a line that slants downward to the right.

Practice

Determine whether a scatter plot of data for the following would be likely to show a *positive*, *negative*, or *no* relationship. Explain your answer.

3. height, shoe size — positive *[handwritten: Because most of the time, the taller you are, the larger your feet are to support your growing weight. ✓]*

4. age, telephone number

5. test grades, minutes spent on homework

6. amount of water in a bathtub, time since the plug was pulled

7. **Standardized Test Practice** What type of relationship would you expect from a scatter plot for data about ages of people under 20 years of age and the number of words in their vocabulary?

    A positive   **B** negative   **C** none

Answers: 1. none 2. negative 3. positive, since the length of the foot tends to increase with increasing height 4. none, since phone numbers are not assigned according to a person's age 5. positive, since studying more tends to make a person better prepared for tests 6. negative, since the amount of water decreases as time increases 7. A

© Glencoe/McGraw-Hill — Glencoe Pre-Algebra

NAME _____ DATE _____ PERIOD ____

1 Chapter Review
Fruity Math

Substitute the values in the box into each problem below and simplify. Write your answer in the blank to the left of the problem.

🍎 = 4   🍊 = −2   🍌 = −1   🍐 = x 🍓 = 6

____9____ 1. (🍌 + 🍎) + 🍓

_____ 2. 🍊(🍓 + 🍌)

_____ 3. 🍊² − 🍎 · (🍊 + 🍓 ÷ 🍌)

_____ 4. 🍊 · 🍐 + 🍌 · 🍐 + 🍎

_____ 5. 🍓(🍐 + 🍌) + 🍌(🍐 − 🍎)

Use the Associative Property of Addition to draw an expression equivalent to the one shown in problem 1.

Answers are located in the Answer Key.

© Glencoe/McGraw-Hill   Glencoe Pre-Algebra

NAME _____ DATE _____ PERIOD _____

2-1 Integers and Absolute Value (Pages 56–61)

An **integer** is a number that is a whole number of units from zero on the number line.

Integers to the left of zero are less than zero. They are **negative**.

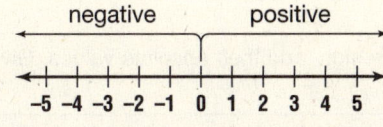

negative ← | → positive
-5 -4 -3 -2 -1 0 1 2 3 4 5
Zero is neither negative nor positive.

Integers to the right of zero are greater than zero. They are **positive**.

The number that corresponds to a point on the number line is the **coordinate** of the point. The **absolute value** of a number is the distance the number is from zero. Two vertical bars are used to indicate the absolute value of a number. For example, $|4| = 4$ and $|-4| = 4$.

Examples

a. Graph {0, −2, 4} on the number line.

Find the point for each number on the number line and draw a dot.

-3 -2 -1 0 1 2 3 4 5

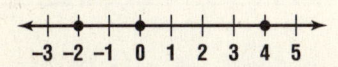

b. Simplify $|-2| + |4|$.

$|-2| + |4| = 2 + 4 = 6$ The absolute value of −2 is 2 and the absolute value of 4 is 4.

Try These Together

Graph each set of numbers on a number line.

1. {1, −1, −3}   2. {5, 7, −3, −2}   3. {4, 2, −2, −4}   4. {2, 3, −4, −3}

HINT: Locate each point on the number line and draw a dot.

Practice

Write an integer for each situation.

5. a loss of $7   −7$
6. a distance of 50 meters
7. 35 minutes left in class

Simplify.

8. $|-12| = 12$
9. $|8|$
10. $|-15| - 8$
11. $|12| + |-2|$
12. $|0| + |-3|$
13. $|-5| - |-2|$
14. $|9| + |-3|$
15. $-|-1|$
16. $|-4| + |3|$
17. $|-6| + |7|$
18. $|8| - |3|$
19. $|14 - 8|$

Evaluate each expression if $a = 4$, $b = 3$ and $c = -2$.

20. $|c| + b$
21. $|a| + |b| - 3$
22. $|b| - |c|$
23. $|a| \cdot |c|$

24. **Standardized Test Practice** An elevator went down 10 floors. What integer describes the trip the elevator made?
A 20   B 10   C −10   D −20

Answers: 1–4. See Answer Key. 5. −7 6. 50 7. 35 8. 12 9. 8 10. 7 11. 14 12. 3 13. 3 14. 12 15. −1 16. 7 17. 13 18. 5 19. 6 20. 5 21. 4 22. 1 23. 8 24. C

© Glencoe/McGraw-Hill   9   Glencoe Pre-Algebra

NAME _____ DATE _____ PERIOD _____

2-2 Adding Integers (Pages 64–68)

You already know that the sum of two positive integers is a positive integer. The rules below will help you find the sign of the sum of two negative integers and the sign of the sum of a positive and a negative integer.

Adding Integers with the Same Sign	To add integers with the same sign, add their absolute values. Give the result the same sign as the integers.
Adding Integers with Different Signs	To add integers with different signs, subtract their absolute values. Give the result the same sign as the integer with the greater absolute value.

Examples

a. Solve $g = -2 + (-10)$.
   Add the absolute values. Give the result the same sign as the integers.
   $g = -(|-2| + |-10|)$
   $g = -(2 + 10)$ or -12

b. Solve $n = -7 + 2$.
   Subtract the absolute values. The result is negative because $|-7| > |2|$.
   $n = -(|-7| - |2|)$
   $n = -(7 - 2)$ or -5

Practice

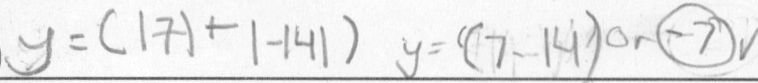

$y = (|7| + |-14|)$ $y = (7-14)$ or (-7)

Solve each equation.

1. $y = 7 + (-14)$
2. $b = -12 + 4$
3. $16 + (-4) = z$
4. $a = 6 + (-15)$
5. $c = 16 + (-15)$
6. $-12 + 31 = q$
7. $-3 + 8 = m$
8. $-4 + 13 = s$
9. $t = (-13) + 7$
10. $-7 + 8 = b$
11. $d = 10 + (-19)$
12. $f = -3 + 17$

Write an addition sentence for each situation. Then find the sum.

13. A hot air balloon is 750 feet high. It descends 325 feet. $z = 750 + (-325)$
14. Cameron owes $800 on his credit card and $750 on his rent. $z = (|750| + |325|)$
   $= (750 + 325)$

Solve each equation.

15. $y = 4 + (-10) + (-2)$
16. $-2 + 4 + (-6) = x$

Simplify each expression.

17. $8x + (-12x)$
18. $-5m + 9m$

19. **Standardized Test Practice** In the high deserts of New Mexico, the morning temperature averages $-2°C$ in the spring. During a spring day, the temperature increases by an average of $27°C$. What is the average high temperature during the spring?
   A $29°C$
   B $25°C$
   C $-25°C$
   D $-29°C$

Answers: 1. -7 2. -8 3. 12 4. -9 5. 1 6. 19 7. 5 8. 9 9. -6 10. 1 11. -9 12. 14 13. $h = 750 + (-325); 425$ 14. $m = -800 + (-750); -1550$ 15. -8 16. -4 17. $-4x$ 18. $4m$ 19. B

© Glencoe/McGraw-Hill   Glencoe Pre-Algebra

NAME _____ DATE _____ PERIOD _____

2-3 Subtracting Integers (Pages 70–74)

Adding and subtracting are inverse operations that "undo" each other. Similarly, when you add opposites, like 4 and −4, they "undo" each other because the sum is zero. An integer and its opposite are called **additive inverses** of each other.

Additive Inverse Property	The sum of an integer and its additive inverse is zero. $5 + (-5) = 0$ or $a + (-a) = 0$

Use the following rule to subtract integers.

Subtracting Integers	To subtract an integer, add its additive inverse. $3 - 5 = 3 + (-5)$ or $a - b = a + (-b)$

Examples

a. Solve $s = -4 - 5$.
  $s = -4 - 5$
  $s = -4 + (-5)$ Add the opposite of 5, or −5.
  $s = -9$

b. Solve $w = 12 - (-6)$.
  $w = 12 - (-6)$
  $w = 12 + 6$ Add the opposite of −6, or 6.
  $w = 18$

Practice

Solve each equation.

1. $x = -3 - 4$ −7
2. $a = -7 - 6$
3. $-18 - 4 = k$
4. $-24 - 7 = b$
5. $-5 - 12 = c$
6. $-18 - 7 = m$
7. $j = 32 - 8$
8. $r = 8 - (-4)$
9. $22 - (-3) = z$
10. $-9 - (-6) = d$
11. $-17 - (-6) = g$
12. $h = 4 - 10$

Evaluate each expression.

13. $n - (-11)$, if $n = 4$
14. $18 - k$, if $k = 5$
15. $9 - (-g)$, if $g = 9$
16. $-11 - k$, if $k = 5$

Simplify each expression.

17. $-x - 7x$
18. $8m - 18m$
19. $-2a - 7a$
20. $9xy - (-8xy)$

21. Is the statement $n = -(-n)$ true or false?

22. **Standardized Test Practice** The elevation of Death Valley, California, is 282 feet below sea level, or −282 feet. To travel from Death Valley to Beatty, Nevada, you must travel over a mountain pass, Daylight Pass, that has an elevation of 4317 feet above sea level. What is the change in elevation from Death Valley to Daylight Pass?

  A 4599 ft   **B** 4035 ft   **C** −4035 ft   **D** −4599 ft

Answers: 1. −7 2. −13 3. −22 4. −31 5. −17 6. −25 7. 24 8. 12 9. 25 10. −3 11. −11 12. −6 13. 15 14. 13 15. 18 16. −16 17. −8x 18. −10m 19. −9a 20. 17xy 21. true 22. A

NAME _____ DATE _____ PERIOD _____

2-4 Multiplying Integers (Pages 75–79)

Use the following rules for multiplying integers.

Multiplying Integers with Different Signs	The product of two integers with different signs is negative.
Multiplying Integers with the Same Signs	The product of two integers with the same sign is positive.

Examples Find the products.

a. $13 \cdot (-12)$
The two integers have different signs. Their product is negative.
$13 \cdot (-12) = -156$

b. $(-15)(-8)$
The two integers have the same sign. Their product is positive.
$(-15)(-8) = 120$

Try These Together

Solve each equation.

1. $y = 8(-12)$
2. $s = -6(9)$
3. $z = (15)(2)$

HINT: Remember, if the factors have the same sign, the product is positive. If the factors have different signs, the product is negative.

Practice

Solve each equation.

4. $-4 \cdot 3 = z$ -12
5. $c = 7(-5)$
6. $d = (-10)(2)$
7. $b = (4)(7)$
8. $t = -6(-2)$
9. $f = (13)(-2)$
10. $g = -10(2)(-3)$
11. $-6(-7)(-2) = a$
12. $14(4)(-1) = h$

Evaluate each expression.

13. $4y$, if $y = -7$
14. gh, if $g = 7$ and $h = -3$
15. $6t$, if $t = 8$
16. $-8d$, if $d = -4$
17. $9xy$, if $x = 2$ and $y = -1$
18. $-3x$, if $x = -13$

Find each product.

19. $7(6x)$
20. $-3gh(-2)$
21. $-14(3d)$
22. $-8x(-2y)$
23. $5n(-7)$
24. $-7(7)(-n)$

25. **Standardized Test Practice** The price of a share of stock changed by $-\$3$ each day for 5 days. What was the overall change in the price of a share of the stock for the 5-day period?

A $15   B $8   C $-\$8$ D $-\$15$

Answers: 1. -96 2. -54 3. 30 4. -12 5. -35 6. -20 7. 28 8. 12 9. -26 10. 60 11. -84 12. -56 13. -28 14. -21 15. 48 16. 32 17. -18 18. 39 19. 42x 20. 6gh 21. -42d 22. 16xy 23. -35n 24. 49n 25. D

© Glencoe/McGraw-Hill   12   Glencoe Pre-Algebra

NAME _____ DATE _____ PERIOD _____

2-5 Dividing Integers (Pages 80–84)

The rules for dividing integers are similar to the rules for multiplying integers.

Dividing Integers with Different Signs	The quotient of two integers with different signs is negative.
Dividing Integers with the Same Signs	The quotient of two integers with the same sign is positive.

Examples Divide.

a. $72 \div (-24)$
The two integers have different signs.
Their quotient is negative.
$72 \div (-24) = -3$

b. $(-65) \div (-5)$
The two integers have the same sign.
Their quotient is positive.
$(-65) \div (-5) = 13$

Practice

Divide.

1. $-48 \div 6$
2. $\dfrac{35}{-7}$
3. $-42 \div -6$
4. $-81 \div 9$
5. $-126 \div (-6)$
6. $36 \div (-3)$
7. $63 \div 9$
8. $-72 \div -9$

9. Divide -48 by 8.
10. Find the quotient of 110 and -11.

Solve each equation.

11. $t = 72 \div -6$
12. $-84 \div 6 = p$
13. $-40 \div (-8) = f$
14. $u = -36 \div (-4)$
15. $128 \div 16 = a$
16. $s = -51 \div (-17)$

Evaluate each expression.

17. $a \div 11$ if $a = -143$
18. $-54 \div (-c)$ if $c = 9$
19. $h \div 12$ if $h = 84$
20. $n \div (-12)$ if $n = -168$
21. $-80 \div k$ if $k = 5$
22. $h \div 7$ if $h = 91$

23. **Weather** The temperature change at a weather station was $-28°F$ in just a few hours. The average hourly change was $-4°F$. Over how many hours did the temperature drop occur?

24. **Standardized Test Practice** Eduardo used money from his savings account to pay back a loan. The change in his balance was $-\$144$ over the period of the loan. What was the monthly change in his balance if he paid back the loan in 3 equal monthly payments?
    A $-\$432$
    B $-\$48$
    C $\$48$
    D $\$432$

Answers: 1. -8 2. -5 3. 7 4. -9 5. 21 6. -12 7. 7 8. 8 9. -6 10. -10 11. -12 12. -14 13. 5 14. 9 15. 8 16. 3 17. -13 18. 6 19. 7 20. 14 21. -16 22. 13 23. 7 hours 24. B

© Glencoe/McGraw-Hill   13   Glencoe Pre-Algebra

NAME _____ DATE _____ PERIOD _____

2-6 The Coordinate System (Pages 85–89)

Coordinate System	A **coordinate system** is formed by two number lines, called **axes**, that intersect at their zero points. The axes separate the coordinate plane into four regions called **quadrants**.

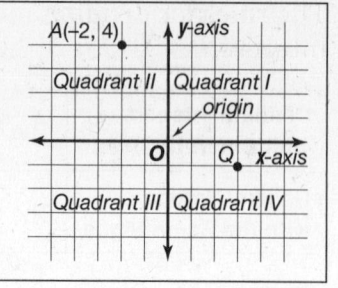

*Any point on the coordinate system is described by an ordered pair, such as (1, −2). In this ordered pair, 1 is the **x-coordinate** and −2 is the **y-coordinate**. If you put a dot on a coordinate system at the point described by (1, −2), you are **plotting the point**. The dot is the **graph** of the point.*

Examples

a. Graph A(−2, 4) on the coordinate system.

Refer to the coordinate system above. Start at the origin. Move 2 units to the left. Then move 4 units up and draw a dot. Label the dot A(−2, 4).

b. What is the ordered pair for point Q on the coordinate system above?

Start at the origin. To get to point Q, move 3 units to the right, and then move 1 unit down. The ordered pair for point Q is (3, −1).

Practice

Name the ordered pair for each point graphed on the coordinate plane.

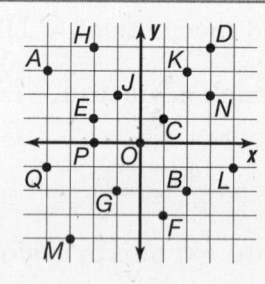

1. H
2. J
3. L
4. G
5. E
6. O
7. B
8. A

What point is located at the following coordinates? Then name the quadrant in which each point is located.

9. (3, 2)
10. (−3, −4)
11. (1, −3)
12. (−2, 0)
13. (−4, −1)
14. (1, 1)
15. (3, 4)
16. (2, 3)

17. **Standardized Test Practice** In a small town, all streets are east-west or north-south. City Center is at (0, 0). City Hall is 1 block north of City Center at (0, 1). City Hospital is 1 block east of City Center at (1, 0). If City Library is 3 blocks north and 2 blocks west of City Center, which ordered pair describes the location of City Library?

A (2, 3)   B (−2, 3)   C (3, −2)   D (3, 2)

Answers: 1. (−2, 4) 2. (−1, 2) 3. (4, −1) 4. (−1, −2) 5. (−2, 1) 6. (0, 0) 7. (2, −2) 8. (−4, 3) 9. N; quadrant I 10. M; quadrant III 11. F; quadrant IV 12. P; no quadrant 13. Q; quadrant III 14. C; quadrant I 15. D; quadrant I 16. K; quadrant I 17. B

© Glencoe/McGraw-Hill   14   Glencoe Pre-Algebra

NAME _____ DATE _____ PERIOD _____

2 Chapter Review

Integer Football

Simplify each expression. Then use your answers to move the team across the football field. Positive answers move the team closer to scoring a touchdown. Negative answers move the team farther *away* from scoring a touchdown. To score a touchdown, the team must cross their opponent's zero-yard (goal) line.

Example

Suppose the team starts on their opponent's 35-yard line.

1st Play: $5 + (-10) =$ __−5__   The team moves back 5 yards to the 40-yard line.

2nd Play: $-2 \cdot (-5) =$ __10__   The team moves forward 10 yards to the 30-yard line.

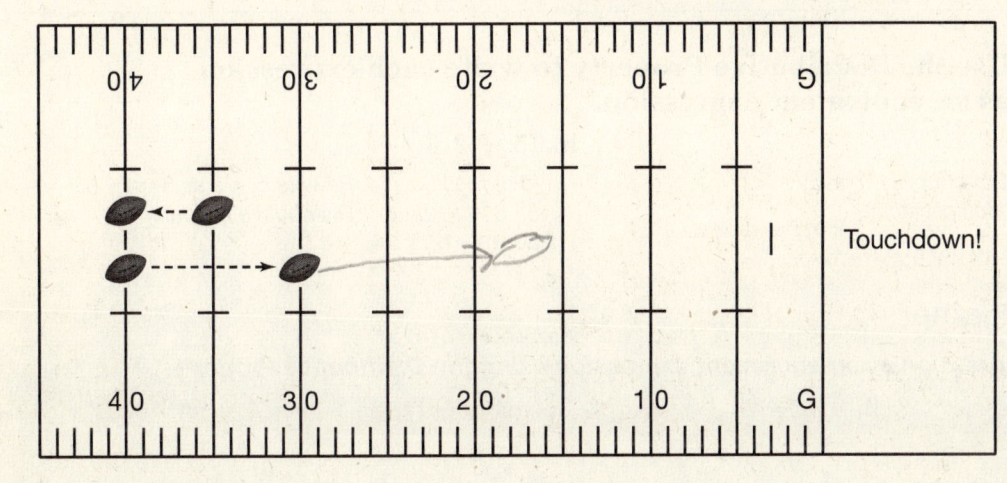

Go!

After an interception, Team A starts on their opponent's 40-yard line.

1st Play: $-36 \div (-3) =$ __12__   What yard line is the team on now? __none__

2nd Play: $20 \div (-4) =$ _____   What yard line is the team on now? _____

3rd Play: $-3 \cdot (-6) =$ _____   What yard line is the team on now? _____

4th Play: $4 - (-12) =$ _____   What yard line is the team on now? _____

Did Team A score a touchdown? Justify your answer.

Answers are located in the Answer Key.

NAME _____ DATE _____ PERIOD _____

3-1 The Distributive Property (Pages 98–102)

The **Distributive Property** allows you to combine addition and multiplication. For example, 5(3 + 1) can be evaluated in two ways. First, we will evaluate 5(3 + 1) by using the order of operations. 5(3 + 1) = 5 · (3 + 1) = 5 · (4) = 20. In this method we added first because the order of operations requires arithmetic within grouping symbols be completed first. Now let's do the same problem by multiplying first.

5(3 + 1) = 5 · (3 + 1) = 5 · 3 + 5 · 1 = 15 + 5 = 20. The second method demonstrates the Distributive Property.

Distributive Property	To multiply a number by a sum, multiply each number in the sum by the number next to the parenthesis. $a(b + c) = ab + ac$ or $(b + c)a = ba + ca$

Examples Use the Distributive Property to write each expression as an equivalent expression.

a. 10(4 + 7)
10 · 4 + 10 · 7   Distributive Property
40 + 70   Multiplication

b. (5 − 2)6
[5 + (−2)]6   Rewrite 5 − 2 as 5 + (−2).
5 · 6 + (−2) · 6   Distributive Property
30 + (−12)

Try These Together

Restate each expression as an equivalent expression using the Distributive Property.

1. 6(7 + 2)   2. 4(9 − 4)   3. −3(5 + 1)   4. −2(8 − 3)

Practice

Use the Distributive Property to write each expression as an equivalent expression. Then evaluate the expression.

5. −2(6 + 1)   6. 13(10 − 7)   7. −11(−3 − 9)   8. [−21 + (−14)]5
9. (7 + 2)4   10. −2(7 − 6)   11. 9(7 + 9)   12. (6 − 3)5

Use the Distributive Property to write each expression as an equivalent algebraic expression.

13. 7(x + 2)   14. 5(b − 8)   15. (q + 9)4   16. 3(c − 6)
17. (m − 2)10 18. −12(d + 14) 19. −18(n − 10) 20. −5(h + 48)

21. **Standardized Test Practice** Use the Distributive Property to write an equivalent algebraic expression for −22(x − y + z − 13).
    A 22x + 22y − 22z + 286
    B −22x − y + z − 13
    C −22x − 22y − 22z − 286
    D −22x + 22y − 22z + 286

Answers: 1. 42 + 12 2. 36 + (−16) 3. −15 + (−3) 4. −16 + 6 5. −14 6. 39 7. 132 8. −175 9. 36 10. −2 11. 144 12. 15 13. 7x + 14 14. 5b + (−40) 15. 4q + 36 16. 3c + (−18) 17. 10m + (−20) 18. −12d + (−168) 19. −18n + 180 20. −5h + (−240) 21. D

© Glencoe/McGraw-Hill   16   Glencoe Pre-Algebra

NAME _____ DATE _____ PERIOD _____

3-2 Simplifying Algebraic Expressions

(Pages 103–107)

An expression such as $5x + 7x$ has two **terms**. These terms are called **like terms** because they have the same variable. You can use the Distributive Property to simplify expressions that have like terms. An expression is in its **simplest form** when it has no like terms and no parentheses.

Distributive Property	The sum of two addends multiplied by a number is the sum of the product of each addend and the number. So, for any numbers a, b, and c, $a(b + c) = ab + ac$ and $(b + c)a = ba + ca$.

Examples   Simplify each expression.

a. $87q + 10q$
   $87q + 10q = (87 + 10)q$ Distributive Property
   $= 97q$

b. $s + 7(s + 1)$
   $s + 7(s + 1) = s + 7s + 7$ Distributive Property
   $= (1 + 7)s + 7$ Distributive Property
   $= 8s + 7$

Try These Together

Restate each expression using the Distributive Property. Do not simplify.

1. $2x + 2y$
2. $n(6 + 4m)$
3. $2(10 + 11)$

Practice

Restate each expression using the Distributive Property. Do not simplify.

4. $z + 6z$ $(1+6)z$
5. $(6 + 10)p$
6. $4t + 8t - 3$
7. $s + 3s + 6s$
8. $4c + 7d + 11d$
9. $2d + 18d$

Simplify each expression.

10. $x + 3x + 10$ $4x + 10$
11. $2x + 4x + 6y$
12. $7(x + 2)$
13. $a + 2b + 7b$
14. $5(6x + 8) + 4x$
15. $y + 2y + 8(y + 7)$

16. **Standardized Test Practice**   Restate the expression $3(x + 2y)$ by using the Distributive Property.
   A $3x + 6y$
   B $3x + 2y$
   C $x + 6y$
   D $6xy$

Answers: 1. $2(x + y)$ 2. $6n + 4mn$ 3. $2(10) + 2(11)$ 4. $(1 + 6)z$ 5. $6p + 10p$ 6. $(4 + 8)t - 3$ 7. $(1 + 3 + 6)s$ 8. $4c + (7 + 11)d$ 9. $(2 + 18)d$ 10. $4x + 10$ 11. $6x + 6y$ 12. $7x + 14$ 13. $a + 9b$ 14. $34x + 40$ 15. $11y + 56$ 16. A

3-3 Solving Equations by Adding or Subtracting (Pages 110–114)

You can use the **Subtraction Property of Equality** and the **Addition Property of Equality** to change an equation into an **equivalent equation** that is easier to solve.

Subtraction Property of Equality	If you subtract the same number from each side of an equation, the two sides remain equal. For any numbers a, b, and c, if a = b, then a − c = b − c.
Addition Property of Equality	If you add the same number to each side of an equation, the two sides remain equal. For any numbers a, b, and c, if a = b, then a + c = b + c.

Examples

a. Solve $q + 12 = 37$.

$q + 12 = 37$
$q + 12 - 12 = 37 - 12$ Subtract 12 from each side.
$q = 25$ Check your solution by replacing q with 25.

b. Solve $k - 23 = 8$.

$k - 23 = 8$
$k - 23 + 23 = 8 + 23$ Add 23 to each side.
$k = 31$ Check your solution.

Practice

Solve each equation and check your solution.

1. $a + 17 = 48$ 48
2. $z + 19 = -4$
3. $b - (-8) = 21$
4. $y + 42 = 103$ 17
5. $129 = g + 59$
6. $39 = h + 14$
7. $c - 17 = 64$ 31 ✓
8. $j + 403 = 564$
9. $64 + r = 108$
10. $s + 18 = 24$
11. $d - (-4) = 52$
12. $78 = f + 61$

Solve each equation. Check each solution.

13. $(18 + y) - 4 = 17$
14. $(p - 4) + 72 = 5$
15. $(n - 11) + 14 = 23$
16. $[k + (-2)] + 18 = 30$
17. $(m + 42) - 23 = 10$
18. $81 = [t - (-4)] + 11$

19. **Sailing** Skip sets sail from Chicago headed toward Milwaukee. Milwaukee is 74 miles from Chicago. He stops for lunch in Kenosha, which is 37 miles from Chicago. How far does he still have to sail?

20. **Standardized Test Practice** In the high mountain plains of Colorado, the temperature can change dramatically during a day, depending upon the Sun and season. On a June day, the low temperature was 14°F. If the high temperature that day was 83°F, by how much had the temperature risen?
    A 50°F   B 69°F   C 70°F   D 83°F

Answers: 1. 31   2. −23   3. 13   4. 61   5. 70   6. 25   7. 81   8. 161   9. 44   10. 6   11. 48   12. 17   13. 3   14. −63   15. 20   16. 14   17. −9   18. 66   19. 37 miles   20. B

NAME _____ DATE _____ PERIOD _____

3-4 Solving Equations by Multiplying or Dividing (Pages 115–119)

Some equations can be solved by multiplying or dividing each side of an equation by the same number.

Division Property of Equality	If you divide each side of an equation by the same nonzero number, the two sides remain equal. For any numbers a, b, and c, where $c \neq 0$ if $a = b$, then $\frac{a}{c} = \frac{b}{c}$.
Multiplication Property of Equality	If you multiply each side of an equation by the same number, the two sides remain equal. For any numbers a, b, and c, if $a = b$, then $a \cdot c = b \cdot c$.

Examples

a. Solve $-6m = 72$.

$-6m = 72$

$\frac{-6m}{-6} = \frac{72}{-6}$ Divide each side by -6.

$m = -12$ Check your solution by replacing m with -12.

b. Solve $\frac{n}{3} = 21$.

$\frac{n}{3} = 21$

$\frac{n}{3} \cdot 3 = 21 \cdot 3$ Multiply each side by 3.

$n = 63$ Check your solution by replacing n with 63.

Try These Together

Solve each equation and check your solution.

1. $36 = 6x$
2. $7b = -49$
3. $\frac{a}{-4} = 6$

Practice

Solve each equation and check your solution.

4. $8c = 72$
5. $-2z = 18$
6. $-42 = 6d$
7. $\frac{m}{12} = 4$
8. $-3h = -36$
9. $\frac{n}{11} = 11$
10. $\frac{s}{-4} = 30$
11. $-524 = -4t$
12. $\frac{k}{6} = 9$
13. $\frac{y}{-18} = -6$
14. $\frac{-x}{9} = -14$
15. $\frac{x}{7} = -20$

16. **Geometry** An equilateral triangle has three sides of equal lengths. If the perimeter of an equilateral triangle is 72 centimeters, how long is each side?

17. **Standardized Test Practice** Enrique has 9 identical bills in his wallet totaling $45.00. What types of bills does he have?

A ones B fives C tens D twenties

Answers: 1. 6 2. −7 3. −24 4. 9 5. −9 6. −7 7. 48 8. 12 9. 121 10. −120 11. 131 12. 54 13. 108 14. 126 15. −140 16. 24 cm 17. B

© Glencoe/McGraw-Hill   Glencoe Pre-Algebra

3-5 Solving Two-Step Equations (Pages 120–124)

To solve an equation with more than one operation, use the work backward strategy and undo each operation. This means you will follow the order of operations in *reverse* order.

Examples Solve each equation. Check your solution.

a. $4a + 12 = 40$

$4a + 12 - 12 = 40 - 12$ Subtract to undo the addition.
$\dfrac{4a}{4} = \dfrac{28}{4}$ Divide to undo the multiplication.
$a = 7$

Does $4(7) + 12 = 40$?
$28 + 12 = 40$
$40 = 40$ True
The solution is 7.

b. $\dfrac{g}{5} - 8 = 7$

$\dfrac{g}{5} - 8 + 8 = 7 + 8$ Add to undo the subtraction.
$\dfrac{g}{5} = 15$
$\dfrac{g}{5} \cdot 5 = 15 \cdot 5$ Multiply to undo the division.
$g = 75$

Does $\dfrac{75}{5} - 8 = 7$?
$15 - 8 = 7$ Do the division first.
$7 = 7$ True
The solution is 75.

Try These Together

Solve each equation. Check your solution.

1. $55 = 4x + 5$
2. $3y - 6 = 3$
3. $4 - 4b = -8$

HINT: Work backward to undo each operation until the variable is alone on one side of the equation.

Practice

Solve each equation. Check your solution.

4. $-5 - 2t = 15$
5. $-4y + 2 = 7$
6. $1.5 = 0.3 + 4y$
7. $14 = 3 + \dfrac{a}{2}$
8. $-\dfrac{3x}{7} = 21$
9. $\dfrac{2}{3}n - 3 = 8$
10. $\dfrac{q - 15}{5} = 4$
11. $\dfrac{6 - x}{4} = -6$
12. $8 = \dfrac{n + 5}{6}$
13. $\dfrac{b}{-3} - 8 = -12$
14. $\dfrac{5 + x}{-12} = -4$
15. $\dfrac{-x - (-3)}{7} = 15$

16. **Consumerism** Carlos bought 5 boxes of floppy disks for his computer. He also bought a paper punch. The paper punch cost $12. The boxes of floppy disks were all the same price. If the total cost before tax was $27, how much did each box of floppy disks cost?

17. **Standardized Test Practice** Solve the equation $\dfrac{-4 - 2x}{9} = 12$.

A −56   B 56   C 112   D 108

Answers: 1. 12.5 2. 3 3. 3 4. −10 5. $-1\dfrac{1}{4}$ 6. 0.3 7. 22 8. −49 9. $16\dfrac{1}{2}$ 10. 35 11. 30 12. 43 13. 12 14. 43 15. −102 16. $3 17. A

© Glencoe/McGraw-Hill   20   Glencoe Pre-Algebra

NAME _____ DATE _____ PERIOD _____

3-6 Writing Two-Step Equations (Pages 126–130)

Many real-world situations can be modeled by two-step equations. In order to find unknown quantities in these situations, you must be able to translate words into equations.

Examples Define a variable and write an equation for each situation. Then solve the equation.

a. Seven less than three times a number is twenty.

Let n represent the number.
Seven less → −7
three times a number → 3n
is twenty → = 20

$3n - 7 = 20$
$3n - 7 + 7 = 20 + 7$ Add 7 to each side.
$\frac{3n}{3} = \frac{27}{3}$ Divide each side by 3.
$n = 9$

b. Four more than a number divided by six is eleven.

Let y represent the number.
Four more → + 4
a number divided by six → $\frac{y}{6}$
is eleven → = 11

$\frac{y}{6} + 4 = 11$
$\frac{y}{6} + 4 - 4 = 11 - 4$ Subtract 4 from each side.
$\frac{y}{6} = 7$
$\frac{y}{6} \cdot 6 = 7 \cdot 6$ Multiply each side by 7.
$y = 42$

Try These Together

Define a variable and write an equation for each situation. Then solve.

1. Three plus 4 times a number is twelve.

2. Six times a number minus five is thirteen.

Practice

Define a variable and write an equation for each situation. Then solve.

3. Two times a number plus eight is eighteen.

4. Twenty-four minus 5 times a number is fifteen.

5. Two times a number minus five is twelve.

6. Six minus the product of four and some number is fifteen.

7. The product of six and some number added to five is fifteen.

8. **Standardized Test Practice** Write an equation for the sentence. The product of some number and five is added to seven to give a total of twenty-three.

   A $x + 5 + 7 = 23$ **B** $x + 5 \div 7 = 23$ **C** $x + 12 = 23$ **D** $5x + 7 = 23$

Answers: 1. $3 + 4x = 12; 2\frac{1}{4}$ 2. $6x - 5 = 13; 3$ 3. $2x + 8 = 18; 5$ 4. $24 - 5x = 15; 1\frac{4}{5}$ 5. $2x - 5 = 12; 8\frac{1}{2}$ 6. $6 - 4x = 15; -2\frac{1}{4}$ 7. $6x + 5 = 15; 1\frac{2}{3}$ 8. D

© Glencoe/McGraw-Hill   21   Glencoe Pre-Algebra

NAME _____ DATE _____ PERIOD _____

3-7 Using Formulas (Pages 131–136)

Formulas can help you solve many different types of problems. A **formula** shows the relationship among certain quantities. For example, to find the number of miles per gallon that a car gets, you can use the following formula: miles driven (m) divided by gallons of gas used (g) equals miles per gallon (mpg), or $m \div g = $ mpg.

Example

Fred bought a sport utility vehicle (SUV), but now he is concerned about the amount of gas it is using. If Fred needs to refill the 25-gallon tank after driving 350 miles, what gas mileage is his SUV getting?

$m \div g = $ mpg	Use the formula.
$350 \div 25 = $ mpg	Replace m with 350 and g with 25.
$350 \div 25 = 14$ mpg	Fred's SUV only gets 14 miles per gallon.

Practice

Solve by replacing the variables in each formula with the given values.

1. $A = \ell w$, if $\ell = 12$ and $w = 9$ A = 12 • 9 = (108) ✓
2. $S = (n - 2)180$, if $n = 4$
3. $I = \frac{1}{20}pt$, if $p = 500$ and $t = 2$
4. $A = \frac{bh}{2}$, if $b = 7$ and $h = 10$
5. $d = 50t$, if $d = 350$
6. $P = 2\ell + 2w$, if $P = 40$ and $\ell = 6$
7. $C = \frac{5}{9}(F - 32)$, if $F = 32$
8. $S = \frac{n(n+1)}{2}$, if $n = 12$
9. **Physics** The density d of a substance is given by the formula $d = \frac{m}{v}$, where m is the mass of a sample of the substance and v is the volume of the sample. Solve $d = \frac{m}{v}$ if $m = 14$ and $v = 2$.
10. **Food** The formula for the circumference of a circle is $C = 2\pi r$, where r is the radius of the circle and π is a constant that is about 3.14. If a pizza has a radius of 8 inches, what is the circumference of the pizza? Round your answer to the nearest inch.
11. **Standardized Test Practice** A train leaves Station A at 11:12 A.M. and arrives at Station B at 2:42 P.M. The train travels at a speed of 80 miles per hour. How many miles does the train travel?
    A 216 mi B 280 mi C 200 mi D 680 mi

Answers: 1. 108 2. 360 3. 50 4. 35 5. 7 6. 14 7. 0 8. 78 9. 7 10. 50 in. 11. B

© Glencoe/McGraw-Hill   22   Glencoe Pre-Algebra

3 Chapter Review

Birthday Puzzle

Today is Mrs. Acevedo's birthday. When her students asked how old she was, she made the following puzzle. For each step of the puzzle, write an equation and solve it. The final step of the puzzle will reveal the year in which she was born. Subtract that year from the current year to find out Mrs. Acevedo's age.

Puzzle

1. The sum of three times a number and 60 is 180. What is the number?

   Student work: $3x + 60 = 180$; $-60\ -60$; $180 - 60 = 120$; $\frac{3x}{3} = \frac{120}{3}$; $x = 40$ ✓

2. Negative one times the answer to problem 1 less five times a number is 210. What is the number?

3. A number divided by eight plus the answer to problem 2 is 100. What is the number?

4. Twice a number less the answer to problem 3 is 6500. What is the number?

5. Four times a number equals 2 times 1925 plus the answer to problem 4. What is the number?

6. Five times a number less 50 is 7900 plus the answer to problem 5. What is the number?

How old is Mrs. Acevedo?

Answers are located in the Answer Key.

© Glencoe/McGraw-Hill — Glencoe Pre-Algebra

4-1 Factors and Monomials (Pages 148–152)

The **factors** of a whole number divide that number with a remainder of 0. For example, 4 is a factor of 12 because 12 ÷ 4 = 3, and 7 is not a factor of 12 because 12 ÷ 7 = 1 with a remainder of 5. Another way of saying that 3 is a factor of 12 is to say that 12 is **divisible** by 3.

Divisibility Rules	A number is divisible by • 2 if the ones digit is divisible by 2. • 3 if the sum of its digits is divisible by 3. • 5 if the ones digit is 0 or 5. • 6 if the number is divisible by 2 and 3. • 10 if the ones digit is 0.

An expression like $5x$ is called a **monomial**. A monomial is an integer, a variable, or a product of integers or variables.

Examples

a. Is $4y(5x)$ a monomial?
Yes, this expression is the product of integers and variables.

b. Is $4y + 5x$ a monomial?
No, this expression is a sum. A sum or difference is not a monomial.

Practice

Using divisibility rules, state whether each number is divisible by 2, 3, 5, 6, or 10.

1. 100   2. 342   3. 600   4. 215
5. 1200   6. 1693   7. 52,700 8. 987,321

Determine whether each expression is a monomial. Explain why or why not.

9. $3x$ 10. -45 11. $2y - 3$ 12. $4(7m)$
13. $x \cdot y \cdot z$ 14. $12 + p$ 15. $2(ab)$ 16. $m + n$

17. **Cake Decorating** If you are decorating a birthday cake using 16 candles, can you arrange all the candles in 6 equal rows? Explain.

18. **Standardized Test Practice** Which of the following is divisible by 3, but is not divisible by 6?
    A 822   B 833   C 922   D 933

Answers: 1. 2, 5, 10 2. 2, 3, 6 3. 2, 3, 5, 6, 10 4. 5 5. 2, 3, 5, 6, 10 6. none 7. 2, 5, 10 8. 3 9. Yes, it is the product of an integer and a variable. 10. Yes, it is an integer. 11. No, it involves subtraction. 12. Yes, it is the product of 4 and 7 times m. 13. Yes, it is the product of x, y, and z. 14. No, it involves addition. 15. Yes, it is the product of 2 and a times b. 16. No, it involves addition. 17. no; 16 is not divisible by 6. 18. D

4-2 Powers and Exponents (Pages 153–157)

An exponent tells how many times a number, called the **base**, is used as a factor. Numbers that are expressed using exponents are called **powers**. Any number, except 0, raised to the zero power, is defined to be 1. So $5^0 = 1$ and $14^0 = 1$. The number 12,345 is in **standard form**. You can use exponents to express a number in expanded form. In **expanded form**, 12,345 is $(1 \times 10^4) + (2 \times 10^3) + (3 \times 10^2) + (4 \times 10^1) + (5 \times 10^0)$.

Powers need to be included in the rules for order of operations.

Order of Operations	1. Do all operations within grouping symbols; start with the innermost grouping symbols. 2. Evaluate all powers in order from left to right. 3. Do all multiplications and divisions in order from left to right. 4. Do all additions and subtractions in order from left to right.

Examples

a. Write $(5 \times 10^3) + (2 \times 10^2) + (7 \times 10^1) + (3 \times 10^0)$ in standard form.
This is $5000 + 200 + 70 + 3$ or 5273.

b. Write 139,567 in expanded form.
$(1 \times 10^5) + (3 \times 10^4) + (9 \times 10^3) + (5 \times 10^2) + (6 \times 10^1) + (7 \times 10^0)$

Try These Together

1. Write (3)(3) using exponents.
HINT: The number 3 is used as a factor 2 times.

2. Write $7 \cdot 7 \cdot 7 \cdot 7 \cdot 7 \cdot 7 \cdot 7$ using exponents. HINT: This is $7^?$.

Practice

Write each multiplication expression using exponents.

3. $a \cdot a \cdot a \cdot a \cdot a$ 4. $(8 \cdot 8)(8 \cdot 8)$ 5. $(x \cdot x)(x \cdot x)(x \cdot x)$ 6. $(-12)(-12)(-12)$

Write each power as a multiplication expression.

7. 14^3 8. m^9 9. $(-2)^4$ 10. y^{10} 11. $(-x)^8$ 12. p^5

Write each number in expanded form.

13. 25   14. 721   15. 1591   16. 40   17. 508   18. 360

19. **Carpeting** Use the formula $A = s^2$ to find how many square feet of carpet are needed to cover a rectangular floor measuring 12 feet by 12 feet.

20. **Standardized Test Practice** Evaluate $m^3 - n^2$ for $m = 3$ and $n = -5$.
A -16 B 2   C 19   D 52

Answers: 1. 3^2 2. 7^7 3. a^5 4. $(8 \cdot 8)^2$, or 8^4 5. $(x \cdot x)^3$, or x^6 6. $(-12)^3$ 7. $14 \cdot 14 \cdot 14$ 8. $m \cdot m \cdot m \cdot m \cdot m \cdot m \cdot m \cdot m \cdot m$ 9. $(-2)(-2)(-2)(-2)$ 10. $y \cdot y \cdot y \cdot y \cdot y \cdot y \cdot y \cdot y \cdot y \cdot y$ 11. $(-x)(-x)(-x)(-x)(-x)(-x)(-x)(-x)$ 12. $p \cdot p \cdot p \cdot p \cdot p$ 13. $(2 \times 10^1) + (5 \times 10^0)$ 14. $(7 \times 10^2) + (2 \times 10^1) + (1 \times 10^0)$ 15. $(1 \times 10^3) + (5 \times 10^2) + (9 \times 10^1) + (1 \times 10^0)$ 16. $(4 \times 10^1) + (0 \times 10^0)$ 17. $(5 \times 10^2) + (0 \times 10^1) + (8 \times 10^0)$ 18. $(3 \times 10^2) + (6 \times 10^1) + (0 \times 10^0)$ 19. 144 ft² 20. B

4-3 Prime Factorization (Pages 159–163)

A **prime number** is a whole number greater than one that has *exactly* two factors, 1 and itself. A **composite number** is a whole number greater than one that has more than two factors. A composite number can always be expressed as a product of two or more primes. When you express a positive integer (other than 1) as a product of factors that are all prime, this is called **prime factorization**. A monomial can be factored as the product of prime numbers, -1, and variables with no exponents greater than 1. For example, $-14cd^2 = -1 \cdot 2 \cdot 7 \cdot c \cdot d \cdot d$.

Finding the Prime Factorization	• The numbers 0 and 1 are neither prime nor composite. • Every number is a factor of 0. The number 1 has only one factor, itself. • Every whole number greater than 1 is either prime or composite. • One way to find the prime factorization of a number is to use a **factor tree** such as the one shown in the Example.

Example

Factor 280 completely.
Use a factor tree like the one shown at the right. The factors are prime. List the prime factors from least to greatest: $280 = 2 \cdot 2 \cdot 2 \cdot 5 \cdot 7$.

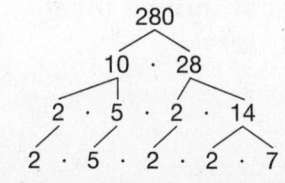

Try These Together

1. Is 13 prime or composite?
2. Is 33 prime or composite?

HINT: You only need to test divisors that are less than half of the number, since a larger divisor would mean that there is also a smaller factor.

Practice

Determine whether each number is prime or composite.

3. 18 composite
4. 37
5. 49
6. 4539

Factor each number or monomial completely.

7. 44
8. 12
9. 90
10. -18
11. -24
12. 28
13. 23
14. -25
15. $8xy^2$
16. $-16ab^3c$
17. $42mn$
18. $50p^2$

19. **Standardized Test Practice** Which of the following is a prime number?
    A 8   B 9   C 13   D 15

Answers: 1. prime 2. composite 3. composite 4. prime 5. composite 6. composite 7. $2 \cdot 2 \cdot 11$ 8. $2 \cdot 2 \cdot 3$ 9. $2 \cdot 3 \cdot 3 \cdot 5$ 10. $-1 \cdot 2 \cdot 3 \cdot 3$ 11. $-1 \cdot 2 \cdot 2 \cdot 2 \cdot 3$ 12. $2 \cdot 2 \cdot 7$ 13. prime 14. $-1 \cdot 5 \cdot 5$ 15. $2 \cdot 2 \cdot 2 \cdot x \cdot y \cdot y$ 16. $-1 \cdot 2 \cdot 2 \cdot 2 \cdot 2 \cdot a \cdot b \cdot b \cdot b \cdot c$ 17. $2 \cdot 3 \cdot 7 \cdot m \cdot n$ 18. $2 \cdot 5 \cdot 5 \cdot p \cdot p$ 19. C

NAME _____ DATE _____ PERIOD ____

4-4 Greatest Common Factor (GCF) *(Pages 164–168)*

The greatest of the factors of two or more numbers is called the **greatest common factor (GCF)**. Two numbers whose GCF is 1 are **relatively prime**.

Finding the GCF	• One way to find the greatest common factor is to list all the factors of each number and identify the greatest of the factors common to the numbers. • Another way is to find the prime factorization of the numbers and then find the product of their common factors.

Examples

a. Find the GCF of 126 and 60.

First find the prime factorization of each number.
$126 = 2 \cdot 3 \cdot 3 \cdot 7$
$60 = 2 \cdot 2 \cdot 3 \cdot 5$
List the common prime factors in each list: 2, 3.
The GCF of 126 and 60 is $2 \cdot 3$ or 6.

b. Find the GCF of $140y^2$ and $84y^3$.

First find the prime factorization of each number.
$140 = 2 \cdot 2 \cdot 5 \cdot 7 \cdot y \cdot y$
$84 = 2 \cdot 2 \cdot 3 \cdot 7 \cdot y \cdot y \cdot y$
List the common prime factors: 2, 2, 7, y, y.
The GCF of $140y^2$ and $84y^3$ is $2 \cdot 2 \cdot 7 \cdot y \cdot y$ or $28y^2$.

Try These Together

1. What is the GCF of 14 and 20?
2. What is the GCF of $21x^4$ and $9x^3$?

HINT: *Find the prime factorization of the numbers and then find the product of their common factors.*

Practice

Find the GCF of each set of numbers or monomials.

3. 6, 18
4. 4, 8, 28
5. 27, 24, 15
6. 6, 10, 25
7. $12x, 3x$
8. $4b, 6ab$
9. $20x, 30y$
10. $14p^2, 28p$
11. $33x^3y, 11x^2y$
12. $30a, 15a^2, 10ab$

Determine whether the numbers in each pair are relatively prime. Write *yes* or *no*.

13. 15 and 12
14. 2 and 9
15. 22 and 21
16. 7 and 63
17. 30 and 5
18. 14 and 35

19. **Quilting** Maria wants to cut two pieces of fabric into the same size squares with no material wasted. One piece measures 12 inches by 36 inches, and the other measures 6 inches by 42 inches. What is the largest size square that she can cut?

20. **Standardized Test Practice** Which of the following is the greatest common factor of 8, 60, and 28?
   A 2
   B 4
   C 60
   D 280

Answers: 1. 2 2. $3x^3$ 3. 6 4. 4 5. 3 6. 1 7. $3x$ 8. $2b$ 9. 10 10. $14p$ 11. $11x^2y$ 12. $5a$ 13. no 14. yes 15. yes 16. no 17. no 18. no 19. 6 in. by 6 in. 20. B

© Glencoe/McGraw-Hill   27   Glencoe Pre-Algebra

4-5 Simplifying Algebraic Fractions *(Pages 169–173)*

A **ratio** is a comparison of two numbers by division. You can express a ratio in several ways. For example, 2 to 3, 2 : 3, $\frac{2}{3}$, and 2 ÷ 3 all represent the same ratio.

Simplifying Fractions	A ratio is most often written as a fraction in **simplest form**. A fraction is in simplest form when the GCF of the numerator and denominator is 1. You can also write **algebraic fractions** that have variables in the numerator or denominator in simplest form.

Examples

a. Write $\frac{8}{12}$ in simplest form.

Find the GCF of 8 and 12.
8 = 2 · 2 · 2
12 = 2 · 2 · 3
The GCF is 2 · 2 or 4.
Divide numerator and denominator by 4.
$\frac{8 \div 4}{12 \div 4} = \frac{2}{3}$

b. Simplify $\frac{15ab^2}{20a^2b}$.

$\frac{15ab^2}{20a^2b} = \frac{3 \cdot 5 \cdot a \cdot b \cdot b}{2 \cdot 2 \cdot 5 \cdot a \cdot a \cdot b}$

Divide numerator and denominator by 5 · a · b.

$\frac{15ab^2}{20a^2b} = \frac{3 \cdot \cancel{5} \cdot \cancel{a} \cdot \cancel{b} \cdot b}{2 \cdot 2 \cdot \cancel{5} \cdot a \cdot \cancel{a} \cdot \cancel{b}}$ or $\frac{3b}{4a}$

Try These Together

1. Write $\frac{8}{16}$ in simplest form.

HINT: Divide the numerator and denominator by the GCF of 8 and 16.

2. Simplify $\frac{6x}{15x^2}$.

HINT: Divide the numerator and denominator by the GCF of 6x and 15x².

Practice

Write each fraction in simplest form. If the fraction is already in simplest form, write *simplified*.

3. $\frac{16}{24}$ **4.** $\frac{10}{45}$ **5.** $\frac{7}{24}$ **6.** $\frac{22}{26}$ **7.** $\frac{12}{21}$ **8.** $\frac{4}{28}$

9. $\frac{40}{50}$ **10.** $\frac{24}{35}$ **11.** $\frac{4x}{8x}$ **12.** $\frac{3m}{27}$ **13.** $\frac{8ab^2}{10ab}$ **14.** $\frac{7x^2}{15x}$

15. Exchange Rates Exchange rates fluctuate daily. Write the ratio of British pounds to American dollars using an exchange rate of £1.00 to $1.60. Simplify your answer.

16. Standardized Test Practice Which of the following is in simplest form?

A $\frac{6}{15}$ **B** $\frac{10}{14}$ **C** $\frac{21}{35}$ **D** $\frac{8}{15}$

Answers: 1. $\frac{1}{2}$ 2. $\frac{2}{5x}$ 3. $\frac{2}{3}$ 4. $\frac{2}{9}$ 5. simplified 6. $\frac{11}{13}$ 7. $\frac{4}{7}$ 8. $\frac{1}{7}$ 9. $\frac{4}{5}$ 10. simplified 11. $\frac{1}{2}$ 12. $\frac{m}{9}$ 13. $\frac{4b}{5}$ 14. $\frac{7x}{15}$ 15. $\frac{5}{8}$ 16. D

© Glencoe/McGraw-Hill   Glencoe Pre-Algebra

4-6 Multiplying and Dividing Monomials (Pages 175–179)

You can multiply and divide numbers with exponents (or powers) if they have the same base.

Multiplying and Dividing Powers	• To find the product of powers *that have the same base*, add their exponents. $a^m \cdot a^n = a^{m+n}$ • To find the quotient of powers *that have the same base*, subtract their exponents. $a^m \div a^n = a^{m-n}$

Examples

a. Find $2^5 \cdot 2^3$.
Follow the pattern of $a^m \cdot a^n = a^{m+n}$. Notice that both factors have the same base, 2. Therefore 2 is also the base of the answer.
$2^5 \cdot 2^3 = 2^{5+3}$ or 2^8

b. Find $\dfrac{b^8}{b^2}$.
Follow the pattern of $a^m \div a^n = a^{m-n}$. Notice that both factors have the same base, b. Therefore the base of the answer is also b.
$\dfrac{b^8}{b^2} = b^{8-2}$ or b^6

Try These Together

1. Find $x \cdot x^3$. Express your answer in exponential form.
   HINT: $x = x^1$

2. Find $\dfrac{9^{10}}{9^6}$. Express your answer in exponential form.
   HINT: The answer will have a base of 9.

Practice

Find each product or quotient. Express your answer in exponential form.

3. $m^4 \cdot m^3$ 4+3=7 (m^7) ✓
4. $(p^{12}q^5)(p^3q^3)$
5. $(2y^7)(5y^2)$
6. $(12x^7)(x^{11})$
7. $8^6 \div 8^2$
8. $\dfrac{15^7}{15^2}$
9. $n^{18} \div n^9$
10. $\dfrac{x^3 y^{10}}{x^3 y^4}$
11. $\dfrac{r^{50}}{r}$
12. $\dfrac{9m^{11}}{3m^5}$
13. $\dfrac{12t^4}{4t^3}$
14. $(x^8 \cdot x^7) \div x^3$

Find each missing exponent.

15. $(y^?)(y^4) = y^{10}$
16. $\dfrac{20^{15}}{20^?} = 20^5$

17. **History** The Italian mathematician Pietro Cataldi, born in 1548, wrote exponents differently from the way they are written today. For example, he wrote 5$\overset{2}{|}$ for $5x^2$ and 5$\overset{3}{|}$ for $5x^3$. How do you think he would have written the answer to $6x^3 \cdot x^4$?

18. **Standardized Test Practice** Simplify the expression $p^6 q^4 r^{10} \cdot p^2 q r^5$.
    A $p^8 q^5 r^{15}$
    B $p^3 q^4 r^2$
    C $p^8 q^4 r^{15}$
    D $p^4 q^3 r^5$

Answers: 1. x^4 2. 9^4 3. m^7 4. $p^{15}q^8$ 5. $10y^9$ 6. $12x^{18}$ 7. 8^4 8. 15^5 9. n^9 10. y^6 11. r^{49} 12. $3m^6$ 13. $3t$ 14. x^{12} 15. 6 16. 10 17. 6$\overset{7}{|}$ 18. A

© Glencoe/McGraw-Hill   Glencoe Pre-Algebra

4-7 Negative Exponents (Pages 181–185)

What does a negative exponent mean? Look at some examples:

$2^{-2} = \frac{1}{2^2}$ or $\frac{1}{4}$ $3^{-4} = \frac{1}{3^4}$ or $\frac{1}{81}$

Negative Exponents	For any nonzero number a and integer n, $a^{-n} = \frac{1}{a^n}$.

Examples

a. Write 2^{-3} using a positive exponent.

$2^{-3} = \frac{1}{2^3}$

b. Write $\frac{1}{3^2}$ as an expression using negative exponents.

$\frac{1}{3^2} = 3^{-2}$

Try These Together

1. Write 7^{-4} using a positive exponent.
   HINT: This is $\frac{1}{7^?}$.

2. Write $\frac{1}{5^2}$ as an expression using negative exponents.
   HINT: The exponent will be -2.

Practice

Write each expression using positive exponents.

3. $x^{-5}y^{-8}$
4. n^{-7}
5. pq^{-2}
6. s^3t^{-2}
7. $a^{-4}b^{-3}c$
8. $\frac{-2x^8}{y^{-9}}$
9. $\frac{(-3)^4}{p^{-10}}$
10. $(-1)^{-3}m^2n^{-1}$
11. $\frac{1}{t^{-7}}$

Write each fraction as an expression using negative exponents.

12. $\frac{1}{2^5}$
13. $\frac{1}{y^6}$
14. $\frac{1}{27}$
15. $\frac{-4}{m^{10}}$
16. $\frac{16}{s^3t^2}$
17. $\frac{a^4}{b^3}$

Evaluate each expression for $n = -2$.

18. n^{-4}
19. 3^n
20. n^{-2}

21. **Physics** The average density of the Earth is about 5.52 grams per cubic centimeter, or $5.52 \text{ g} \cdot \text{cm}^{-3}$. Write this measurement as a fraction using positive exponents.

22. **Standardized Test Practice** Express $a^3b^{-4}c^2d^{-1}$ using positive exponents.

A $\frac{a^3b^4}{c^2d}$ B $a^3b^4c^2d$ C $\frac{b^4d}{a^3c^2}$ D $\frac{a^3c^2}{b^4d}$

Answers: 1. $\frac{1}{7^4}$ 2. 5^{-2} 3. $\frac{1}{x^5y^8}$ 4. $\frac{1}{n^7}$ 5. $\frac{p}{q^2}$ 6. $\frac{s^3}{t^2}$ 7. $\frac{c}{a^4b^3}$ 8. $-2x^8y^9$ 9. $(-3)^4p^{10}$ 10. $\frac{m^2}{(-1)^3n}$ 11. t^7 12. 2^{-5} 13. y^{-6} 14. 3^{-3} 15. $-4m^{-10}$ 16. $16s^{-3}t^{-2}$ 17. $\frac{b^{-3}}{a^{-4}}$ 18. $\frac{1}{16}$ 19. $\frac{1}{9}$ 20. $\frac{1}{4}$ 21. $\frac{5.52 \text{ g}}{\text{cm}^3}$ 22. D

NAME _____ DATE _____ PERIOD _____

4-8 Scientific Notation (Pages 186–190)

You can use **scientific notation** to write very large or very small numbers. Numbers expressed in scientific notation are written as the product of a factor and a power of 10. The factor must be greater than or equal to 1 and less than 10.

Scientific Notation	• To write a large positive or negative number in scientific notation, move the decimal point to the right of the left-most digit, and multiply this number by a power of ten. • To find the power of ten, count the number of places you moved the decimal point. • The procedure is the same for small numbers, except the power of 10 is the negative of the number of places you moved the decimal point.

Examples Write each number in scientific notation.

a. 93,000,000

9.3000000.   Move the decimal point 7 spaces to the left.

9.3×10^7 Multiply by a factor of 10, which in this case is 10^7 because you moved the decimal point 7 spaces to the left.

b. 0.0000622

0.0006.22   Move the decimal point 5 spaces to the right.

6.22×10^{-5} Multiply by 10^{-5} because you moved the decimal point 5 spaces to the right.

c. Write 8.3×10^{-4} in standard form.

$8.3 \times 10^{-4} = 8.3 \times \left(\frac{1}{10}\right)^4$

$= 8.3 \times \frac{1}{10,000}$

$= 8.3 \times 0.0001$ or 0.00083 Move the decimal point in 8.3 4 places to the left.

Practice

Write each number in scientific notation.

1. 3,265,000 (3.265 · 10⁶)
2. 4,560,000
3. 5,200,000,000
4. 0.00057
5. 0.00000002
6. 73,000,000,000

Write each number in standard form.

7. 5.7×10^6
8. 6.8×10^8
9. 3.2×10^{-5}
10. 6.7×10^{-7}
11. 5.9×10^{12}
12. 3.034579×10^6

13. **Chemistry** Because atoms are so small, chemists use metric prefixes to describe extremely small numbers. A *femtogram* is 0.000000000000001 of a gram. Write this number in scientific notation.

14. **Standardized Test Practice** Write 640,000,000, in scientific notation.
    A 6.4×10^8
    B 6.4×10^{11}
    C 6.4×10^{-8}
    D 64×10^{-11}

Answers: 1. 3.265×10^6 2. 4.56×10^6 3. 5.2×10^9 4. 5.7×10^{-4} 5. 2.0×10^{-8} 6. 7.3×10^{10} 7. 5,700,000 8. 680,000,000 9. 0.000032 10. 0.00000067 11. 5,900,000,000,000 12. 3,034,579 13. 1.0×10^{-15} 14. A

© Glencoe/McGraw-Hill   Glencoe Pre-Algebra

NAME _____ DATE _____ PERIOD _____

4 Chapter Review

Puzzling Factors and Fractions

Use the following clues to complete the puzzle at the right. Here are a few examples of how exponents and fractions should be entered into the puzzle.

$\frac{5}{23}$ ⇒ | 5 | 2 | 3 |

$7x^3y^4$ ⇒ | 7 | x | 3 | y | 4 |

$\frac{3}{4a^2}$ ⇒ | 3 | 4 | a | 2 |

(crossword grid with 1-Across filled in as "6ab3")

ACROSS

1. The quotient $\frac{24ab^5}{4b^2}$ 6ab³ ✓

3. $\frac{8a^2b}{4ab^3}$ in simplified form

4. The value of 3^{-n} if $n = 4$

5. $\frac{36}{63}$ in simplest form

8. $5xy^{-3}$ written using positive exponents

10. The product of 2^3 and 7

12. $\frac{15x^5y^2}{90xy^3}$ in simplest form

13. The product $(3m)(16n)$

15. The GCF of 60 and 90

DOWN

1. The product of $12a$ and $5a^3$

2. The value of $a^2 - b$ if $a = -5$ and $b = 3$

3. The product $(7xy^3)(3x^2y)$

6. $\frac{1}{7^{-5}}$ written using positive exponents

7. The GCF of 30 and 45

9. The quotient $\frac{x^3y^5}{xy^2}$

11. The GCF of $42mn^3$ and $54m^2n$

12. The product of x^4 and x^2

14. The quotient of 8^7 and 8^4

Answers are located in the Answer Key.

© Glencoe/McGraw-Hill   32   Glencoe Pre-Algebra

NAME _____ DATE _____ PERIOD _____

5-1 Writing Fractions as Decimals *(Pages 200–204)*

To change a fraction to an **equivalent decimal**, divide the numerator by the denominator. If the division comes to an end (that is, gives a remainder of zero), the decimal is a *terminating* decimal. If the division never ends (that is, never gives a zero remainder), the decimal is a *repeating* decimal. For example, $\frac{1}{8}$ gives the terminating decimal 0.125, and $\frac{5}{6}$ gives the repeating decimal 0.8333..., which is written $0.8\overline{3}$. The bar over the 3 indicates that the 3 repeats forever. You can use a calculator to change a fraction to a decimal.

Examples

a. Write $2\frac{2}{5}$ as a decimal.

Method 1: Use paper and pencil.

$2\frac{2}{5} = 2 + \frac{2}{5}$

$.4$
$5\overline{)2.0}$
$\underline{-2\,0}$
0

So $2 + 0.4 = 2.4$.

Method 2: Use a calculator.
Enter 2 [+] 2 [÷] 5 [=]. Result: 2.4.
Make sure your calculator follows the order of operations.

b. Replace ● with <, >, or =: $\frac{2}{3}$ ● $\frac{3}{4}$.

Method 1: Rewrite as decimals.

$\frac{2}{3} = 0.\overline{6}$ $\qquad \frac{3}{4} = 0.75$

$0.6 < 0.75$

Method 2: Write equivalent fractions with like denominators.

The LCM is 12.

$\frac{2}{3} = \frac{8}{12}$ and $\frac{3}{4} = \frac{9}{12}$

$\frac{8}{12} < \frac{9}{12}$, so $\frac{2}{3} < \frac{3}{4}$.

Try These Together

Write each fraction as a decimal. Use a bar to show a repeating decimal.

1. $\frac{4}{10}$
2. $\frac{7}{9}$
3. $-\frac{1}{2}$
4. $5\frac{7}{16}$

Practice

Write each fraction as a decimal. Use a bar to show a repeating decimal.

5. $-\frac{3}{4}$ —0.75 ✓
6. $4\frac{16}{20}$
7. $\frac{3}{9}$
8. $\frac{18}{25}$

Replace each ● with >, <, or = to make a true sentence.

9. $\frac{7}{8}$ ● $\frac{5}{9}$
10. $-2\frac{2}{5}$ ● $-2\frac{1}{4}$
11. $\frac{7}{12}$ ● $\frac{21}{36}$

12. **Standardized Test Practice** An airplane flies at about 600 miles per hour. At some point during its landing, it drops to about $\frac{2}{9}$ of this speed. Write this fraction as a decimal.

A 0.60 B 0.50 C 0.40 D $0.\overline{2}$

Answers: 1. 0.4 2. $0.\overline{7}$ 3. −0.5 4. 5.4375 5. −0.75 6. 4.8 7. $0.\overline{3}$ 8. 0.72 9. > 10. < 11. = 12. D

© Glencoe/McGraw-Hill — Glencoe Pre-Algebra

NAME _____ DATE _____ PERIOD _____

5-2 Rational Numbers (Pages 205–209)

Sets of Numbers	• The set of **whole numbers** is {0, 1, 2, 3, 4, 5, …}. Such numbers as $\frac{5}{5}$, $\frac{9}{1}$, and $\frac{25}{5}$ are also whole numbers because they can be written as a member of this set. • The set of **integers** is the set of whole numbers and their opposites. • The set of **rational numbers** consists of all numbers that can be expressed as $\frac{a}{b}$, where a and b are integers and $b \neq 0$. The numbers $\frac{1}{3}$ and -5 are rational numbers.

Some decimals are rational numbers.

Types of Decimals	• Decimals either terminate (come to an end) or they go on forever. Every **terminating decimal** can be written as a fraction, so all terminating decimals are rational numbers. For example, $0.45 = \frac{45}{100}$ or $\frac{9}{20}$. • **Repeating decimals** can always be written as fractions, so repeating decimals are always rational numbers. You can use **bar notation** to indicate that some part of a decimal repeats forever, for example, $0.333… = 0.\overline{3}$. • Decimals that do not terminate and do not repeat cannot be written as fractions and are not rational numbers.

Example

Express $0.\overline{23}$ as a fraction in simplest form.

Let $N = 0.232323…$.   Then $100N = 23.232323…$.

$100N = 23.232323…$ Multiply N by 100 because two digits repeat.
$-\ N = 0.232323…$ Subtract N from 100N to eliminate the repeating part.
$99N = 23$ ⇒ $N = \frac{23}{99}$ To check this answer divide 23 by 99.

Practice

Express each decimal as a fraction or mixed number in simplest form.

1. 0.6
2. 0.444…
3. $-0.\overline{15}$
4. 1.26

Name the set(s) of numbers to which each number belongs.

5. $\frac{3}{8}$
6. -1280
7. -2.5
8. $-0.\overline{53}$

Replace each ● with <, >, or = to make a true sentence.

9. $\frac{1}{3}$ ● $0.\overline{3}$
10. -2 ● 2.25
11. 1.8 ● $1.\overline{7}$
12. $\frac{6}{8}$ ● 0.75

13. **Standardized Test Practice** Which number is the greatest, $\frac{5}{10}$, $\frac{6}{11}$, $\frac{6}{13}$, or $\frac{4}{9}$?

A $\frac{4}{9}$ B $\frac{6}{11}$ C $\frac{5}{10}$ D $\frac{6}{13}$

Answers: 1. $\frac{3}{5}$ 2. $\frac{4}{9}$ 3. $-\frac{5}{33}$ 4. $1\frac{13}{50}$ 5. rational 6. integer, rational 7. rational 8. rational 9. = 10. > 11. < 12. = 13. B

NAME _____ DATE _____ PERIOD _____

5-3 Multiplying Rational Numbers (Pages 210–214)

Multiplying Fractions	To multiply fractions, multiply the numerators and multiply the denominators. For fractions $\frac{a}{b}$ and $\frac{c}{d}$, where $b \neq 0$ and $d \neq 0$, $\frac{a}{b} \cdot \frac{c}{d} = \frac{ac}{bd}$. If fractions have common factors in the numerators and denominators, you can simplify before you multiply.

Examples

a. Solve $x = \frac{1}{5} \cdot \frac{2}{3}$.

$x = \frac{1}{5} \cdot \frac{2}{3}$

$= \frac{1 \cdot 2}{5 \cdot 3}$ or $\frac{2}{15}$

b. Solve $y = \frac{3}{4} \cdot \frac{2}{5}$.

$y = \frac{3}{4} \cdot \frac{2}{5}$

$= \frac{3 \cdot \cancel{2}^{1}}{\cancel{4}_{2} \cdot 5}$ The GCF of 2 and 4 is 2. Divide 2 and 4 by 2.

$= \frac{3 \cdot 1}{2 \cdot 5}$ or $\frac{3}{10}$

Try These Together

Solve each equation. Write the solution in simplest form.

1. $t = \frac{2}{3} \cdot \frac{1}{4}$

2. $\left(\frac{3}{5}\right)\left(\frac{1}{2}\right) = g$

3. $c = \left(\frac{3}{5}\right)\left(-\frac{1}{4}\right)$

Practice

Solve each equation. Write the solution in simplest form.

4. $\left(-\frac{9}{10}\right)(-3) = h$

5. $-\frac{1}{2} \cdot \left(\frac{3}{4}\right) = d$

6. $m = 18\left(-\frac{2}{3}\right)$

7. $5\left(-\frac{12}{15}\right) = a$

8. $n = \left(-\frac{5}{3}\right)\left(\frac{4}{2}\right)$

9. $\left(-\frac{11}{20}\right) \cdot 4 = k$

10. $p = 3\left(-\frac{3}{3}\right)$

11. $\left(-\frac{15}{21}\right)\left(-\frac{3}{5}\right) = w$

12. $r = \left(-\frac{6}{18}\right)\left(\frac{9}{12}\right)$

13. What is the product of $\frac{12}{20}$ and $\frac{2}{3}$?

14. What is $\frac{5}{8}$ of 42?

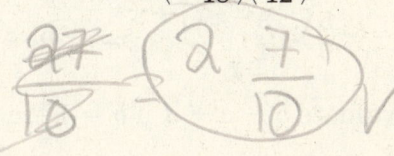

15. Standardized Test Practice Jemeal has $75 to go shopping. She spends $\frac{1}{3}$ of her money on CDs and $\frac{1}{8}$ on food at the food court. About how much money does she have left?

A $54 **B** $41 **C** $33 **D** $24

Answers: 1. $\frac{1}{6}$ 2. $\frac{3}{10}$ 3. $-\frac{3}{20}$ 4. $\frac{27}{10}$ 5. $-\frac{3}{8}$ 6. -12 7. -4 8. $-\frac{10}{3}$ 9. $-\frac{11}{5}$ 10. -3 11. $\frac{3}{7}$ 12. $-\frac{1}{4}$ 13. $\frac{2}{5}$ 14. $26\frac{1}{4}$ 15. B

© Glencoe/McGraw-Hill   35   Glencoe Pre-Algebra

NAME _____ DATE _____ PERIOD _____

5-4 Dividing Rational Numbers *(Pages 215–219)*

Two numbers whose product is 1 are **multiplicative inverses**, or **reciprocals** of each other. For example, 2 and $\frac{1}{2}$ are reciprocals of each other since $2 \times \frac{1}{2} = 1$.

Inverse Property of Multiplication	For every nonzero number $\frac{a}{b}$ where $a, b \neq 0$, there is exactly one number $\frac{b}{a}$ such that $\frac{a}{b} \cdot \frac{b}{a} = 1$.
Division with Fractions	To divide by a fraction, multiply by its multiplicative inverse. For fractions $\frac{a}{b}$ and $\frac{c}{d}$, where $b, c,$ and $d \neq 0$, $\frac{a}{b} \div \frac{c}{d} = \frac{a}{b} \cdot \frac{d}{c}$.

Examples

a. Solve $d = \frac{1}{2} \div \frac{7}{8}$.

$d = \frac{1}{2} \div \frac{7}{8}$

$= \frac{1}{2} \cdot \frac{8}{7}$ $\frac{8}{7}$ is the multiplicative inverse of $\frac{7}{8}$.

$= \frac{1 \cdot \cancel{8}^4}{\cancel{2} \cdot 7}$ or $\frac{4}{7}$

b. Solve $g = \frac{5}{6} \div 1\frac{1}{2}$.

$g = \frac{5}{6} \div 1\frac{1}{2}$

$= \frac{5}{6} \div \frac{3}{2}$ Rename $1\frac{1}{2}$ as $\frac{3}{2}$.

$= \frac{5}{6} \cdot \frac{2}{3}$ $\frac{2}{3}$ is the multiplicative inverse of $\frac{3}{2}$.

$= \frac{5 \cdot \cancel{2}^1}{\cancel{6}_3 \cdot 3}$ or $\frac{5}{9}$

Practice

Estimate the solution to each equation. Then solve. Write the solution in simplest form.

1. $p = \frac{6}{10} \div \left(-\frac{5}{8}\right)$

2. $-\frac{19}{21} \div \left(-\frac{3}{7}\right) = w$

3. $r = -\frac{4}{8} \div \frac{9}{16}$

4. $k = -\frac{5}{6} \div \frac{3}{4}$

5. $s = -\frac{8}{9} \div \left(-\frac{8}{18}\right)$

6. $7 \div \left(-\frac{8}{10}\right) = b$

7. Evaluate $b - c \div d$ if $b = 1\frac{4}{5}$, $c = 1\frac{1}{3}$, and $d = \frac{5}{8}$.

8. **Pets** Students at Midtown Middle School decided to make and donate dog leashes to the local animal shelter. They had 150 meters of leash rope. Each leash was to be $1\frac{2}{3}$ meters long. How many leashes can the students make?

9. **Standardized Test Practice** Solve $q = \frac{5}{6} \div 1\frac{2}{3}$. Write the solution in simplest form.

   A $\frac{1}{2}$ B $\frac{18}{25}$ C $1\frac{7}{18}$ D 2

Answers: 1. $-\frac{24}{25}$ 2. $2\frac{1}{9}$ 3. $-\frac{8}{9}$ 4. $-\frac{10}{9}$ 5. 2 6. $-8\frac{3}{4}$ 7. $-\frac{1}{3}$ 8. 90 leashes 9. A

© Glencoe/McGraw-Hill   36   Glencoe Pre-Algebra

NAME _____ DATE _____ PERIOD _____

5-5 Adding and Subtracting Like Fractions (Pages 220–224)

You can add or subtract fractions when they have the same denominators (or *like* denominators). When the sum of two fractions is greater than one, you usually write the sum as a mixed number in simplest form. A **mixed number** indicates the sum of a whole number and a fraction.

Adding and Subtracting Like Fractions	To add or subtract fractions with like denominators, add or subtract the numerators and write the sum over the same denominator. $\frac{a}{c} + \frac{b}{c} = \frac{a+b}{c}$ and $\frac{a}{c} - \frac{b}{c} = \frac{a-b}{c}$, where $c \neq 0$.

Examples

a. Solve $r = 1\frac{2}{3} + 4\frac{1}{3}$.

$r = (1 + 4) + \left(\frac{2}{3} + \frac{1}{3}\right)$ Add the whole numbers and fractions separately.

$r = 5 + \frac{3}{3}$

$r = 5 + 1$ or 6 $\frac{3}{3} = 1$

b. Solve $g = \frac{14}{15} - \frac{30}{15}$.

$g = \frac{14 - 30}{15}$ Subtract the numerators.

$g = -\frac{16}{15}$

$g = -\frac{15}{15} + \left(-\frac{1}{15}\right)$ or $-1\frac{1}{15}$ Rewrite as a mixed number.

Try These Together

1. Solve $k = 6\frac{4}{5} - 2\frac{1}{5}$ and write the solution in simplest form.

2. Solve $\frac{3}{10} + \frac{7}{10} = n$ and write the solution in simplest form.

Practice

Solve each equation. Write the solution in simplest form.

3. $\frac{15}{18} - \frac{10}{18} = t$

4. $x = \frac{13}{21} + \frac{10}{21}$

5. $r = -\frac{4}{35} + \frac{9}{35}$

6. $m = 2\frac{5}{7} + 1\frac{3}{7}$

7. $2\frac{1}{9} - \frac{8}{9} = p$

8. $j = 4\frac{2}{3} + 7\frac{1}{3}$

9. $q = 1\frac{5}{16} - \frac{10}{16}$

10. $w = 2\frac{16}{21} + \left(-\frac{2}{21}\right)$

11. $\frac{3}{8} - \left(-1\frac{1}{8}\right) = b$

12. Simplify the expression $\frac{2}{3}x + \frac{1}{3}x + 2\frac{1}{3}x$.

13. **Standardized Test Practice** Evaluate the expression $x - y$ for $x = \frac{7}{9}$ and $y = \frac{1}{9}$.

A $\frac{8}{9}$ B $\frac{2}{3}$ C $\frac{5}{9}$ D $\frac{1}{3}$

Answers: 1. $4\frac{3}{5}$ 2. 1 3. $\frac{5}{18}$ 4. $1\frac{2}{21}$ 5. $\frac{1}{7}$ 6. $4\frac{1}{7}$ 7. $1\frac{2}{9}$ 8. 12 9. $\frac{11}{16}$ 10. $2\frac{2}{3}$ 11. $1\frac{1}{2}$ 12. $3\frac{1}{3}x$ 13. B

NAME _____ DATE _____ PERIOD _____

5-6 Least Common Multiple (LCM) *(Pages 226–230)*

A **multiple** of a number is a product of that number and any whole number. Multiples that are shared by two or more numbers are called **common multiples**. The least nonzero common multiple of two or more numbers is called the **least common multiple (LCM)** of the numbers.

Comparing Fractions	One way to compare fractions is to write equivalent fractions with the *same* denominator. The most convenient denominator to use is usually the least common multiple of the denominators, or the **least common denominator (LCD)** of the fractions.

Examples

a. Find the LCM of $6a^2$ and $9a$.

Find the prime factorization of each monomial.
$6a^2 = 2 \cdot 3 \cdot a \cdot a$
$9a = 3 \cdot 3 \cdot a$
Find the common factors. Then multiply all of the factors, using the common factors only once.
$2 \cdot 3 \cdot 3 \cdot a \cdot a = 18a^2$
So the LCM of $6a^2$ and $9a$ is $18a^2$.

b. Compare $\frac{11}{12}$ and $\frac{13}{16}$.

$12 = 2 \cdot 2 \cdot 3$ and $16 = 2 \cdot 2 \cdot 2 \cdot 2$, so the LCM of the denominators, or LCD, is $2 \cdot 2 \cdot 2 \cdot 2 \cdot 3$ or 48.
Find equivalent fractions with 48 as the denominator.
$\frac{11 \times 4}{12 \times 4} = \frac{44}{48}$ $\frac{13 \times 3}{16 \times 3} = \frac{39}{48}$
Since $\frac{44}{48} > \frac{39}{48}, \frac{11}{12} > \frac{13}{16}$.

Try These Together

1. Find the LCM of $8x$ and $6y$.

   HINT: Begin by finding the prime factorization of each number.

2. Compare $\frac{4}{7}$ and $\frac{2}{3}$.

   HINT: Write equivalent fractions using the LCM of 7 and 3.

Practice

Find the LCM of each set of numbers or algebraic expressions.

3. 10, 2
4. 14, 4
5. $2b, 8b$
6. $12t, 8t$
7. $22m, 11n$
8. 5, 4, 3
9. $15a^2, 3a^3$
10. $2x, 10xy, 3z$

First find the LCD for each pair of fractions. Then replace the ● with <, >, or = to make a true statement.

11. $\frac{3}{4}$ ● $\frac{5}{8}$
12. $\frac{1}{10}$ ● $\frac{2}{12}$
13. $\frac{6}{7}$ ● $\frac{4}{5}$
14. $\frac{5}{9}$ ● $\frac{11}{21}$

15. **Standardized Test Practice** What is the LCM of 2, 8, and 6?
   A 2
   B 14
   C 24
   D 48

Answers: 1. $24xy$ 2. $\frac{4}{7} < \frac{2}{3}$ 3. 10 4. 28 5. $8b$ 6. $24t$ 7. $22mn$ 8. 60 9. $15a^3$ 10. $30xyz$ 11. $8; <$ 12. $60; >$ 13. $35; >$ 14. $63; >$ 15. C

© Glencoe/McGraw-Hill   38   Glencoe Pre-Algebra

5-7 Adding and Subtracting Unlike Fractions (Pages 232–236)

You can add or subtract fractions with unlike denominators by renaming them with a common denominator. One way to rename unlike fractions is to use the LCD (least common denominator).

Examples

a. Solve $a = 2\frac{3}{4} + 5\frac{2}{3}$.

$a = 2\frac{3}{4} \cdot \frac{3}{3} + 5\frac{2}{3} \cdot \frac{4}{4}$ The LCD is $2 \cdot 2 \cdot 3$ or 12.

$a = 2\frac{9}{12} + 5\frac{8}{12}$ Rename each fraction with the LCD.

$a = 7\frac{17}{12}$ Add the whole numbers and then the like fractions.

$a = 7 + 1\frac{5}{12}$ or $8\frac{5}{12}$ Rename $\frac{17}{12}$ as $1\frac{5}{12}$.

b. Solve $x = 8\frac{2}{5} - 2\frac{9}{10}$.

$x = 8\frac{4}{10} - 2\frac{9}{10}$ The LCD is 10. Rename the fractions.

$x = 7\frac{14}{10} - 2\frac{9}{10}$ Rename $8\frac{4}{10}$ as $7 + 1\frac{4}{10}$ or $7\frac{14}{10}$.

$x = 5\frac{5}{10}$ or $5\frac{1}{2}$ Subtract and simplify.

Try These Together

1. Solve $a = \frac{2}{3} + \frac{1}{12}$. Write the solution in simplest form.
HINT: The LCD of 3 and 12 is 12.

2. Solve $x = \frac{5}{8} - \frac{1}{3}$. Write the solution in simplest form.
HINT: The LCD of 8 and 3 is 24.

Practice

Solve each equation. Write the solution in simplest form.

3. $y = \frac{13}{21} - \frac{1}{3}$

4. $\frac{3}{20} - \frac{1}{2} = n$

5. $c = \frac{11}{15} + \frac{2}{5}$

6. $1\frac{1}{6} - \frac{1}{2} = p$

7. $g = 3\frac{4}{5} + 1\frac{1}{10}$

8. $8\frac{2}{9} - \frac{1}{3} = d$

9. $m = \frac{1}{2} + \frac{3}{5}$

10. $\frac{2}{3} - \frac{1}{2} = q$

11. $t = \frac{5}{6} - \frac{3}{10}$

12. $1\frac{1}{2} + 2\frac{1}{6} = j$

13. $3\frac{2}{5} - 2\frac{1}{6} = w$

14. $h = \frac{3}{50} + \frac{2}{25}$

Evaluate each expression if $x = \frac{1}{2}$, $y = -\frac{2}{3}$, and $z = \frac{3}{4}$. Write in simplest form.

15. $z - x$

16. $x + y + z$

17. $x - y - z$

18. **Standardized Test Practice** Simplify the expression $\frac{3}{8} + \frac{1}{2} + \frac{1}{2}$.

   A $1\frac{5}{8}$ **B** $1\frac{3}{8}$ **C** $\frac{9}{8}$ **D** $\frac{7}{8}$

Answers: 1. $\frac{3}{4}$ 2. $\frac{7}{24}$ 3. $\frac{2}{7}$ 4. $-\frac{7}{20}$ 5. $1\frac{2}{15}$ 6. $\frac{2}{3}$ 7. $4\frac{9}{10}$ 8. $7\frac{8}{9}$ 9. $1\frac{1}{10}$ 10. $\frac{1}{6}$ 11. $\frac{8}{15}$ 12. $3\frac{2}{3}$ 13. $1\frac{7}{30}$ 14. $\frac{7}{50}$ 15. $\frac{1}{4}$ 16. $\frac{7}{12}$ 17. $\frac{5}{12}$ 18. B

© Glencoe/McGraw-Hill   Glencoe Pre-Algebra

NAME _____ DATE _____ PERIOD _____

5-8 Measures of Central Tendency (Pages 238–242)

To analyze sets of data, researchers often try to find a number or data item that can represent the whole set. These numbers or pieces of data are called **measures of central tendency**.

Mean	The **mean** of a set of data is the sum of the data divided by the number of pieces of data. The mean is the same as the *arithmetic average* of the data.
Median	The **median** is the number in the middle when the data are arranged in order. When there are two middle numbers, the median is their mean.
Mode	The **mode** of a set of data is the number or item that appears most often. If no data item occurs more often than others, there is *no mode*.

Example

Find the mean, median, and mode of the following data set.
80, 90, 85, 80, 90, 90, 40, 85

To find the mean, find the sum of the data, divided by the number of pieces of data, or 8.

$$\frac{80 + 90 + 85 + 80 + 90 + 90 + 40 + 85}{8}$$

mean = 80

To find the median, first put the data set in order from least to greatest.
45, 80, 80, 85, 85, 90, 90, 90
The median is the mean of the middle two items, or $\frac{85 + 85}{2}$.
median = 85

The mode is the number of items that occur most often. 90 occurs three times, which is the most often of any data number.
mode = 90

Practice

Find the mean, median and mode for each set of data. When necessary, round to the nearest tenth.

1. 18, 23, 7, 33, 26, 23, 42, 18, 11, 25, 23
2. 25, 26, 27, 28, 28, 29, 30, 31, 30, 29, 28, 27, 26, 25
3. 103, 99, 114, 22, 108, 117, 105, 100, 96, 99, 119
4. 2.3, 5.6, 3.4, 7.3, 6.5, 2.9, 7.7, 8.1, 4.6, 2.3, 8.5

5. **School Populations** The table at the right shows the size of each ethnic group in the Central School District student population. Find the mean, median, and mode for the data set.

Ethnic Group	Number of Students
Asian American	534
African American	678
European American	623
Hispanic American	594
Native American	494

6. **Standardized Test Practice** What is the mean of this data set?
1, 2, 3, 4, 5, 6, 5, 4, 3, 2, 1
A 36   B 6   C 3.27   D 3.0

Answers: 1. mean = 22.6, median = 23, mode = 23 2. mean = 27.8, median = 28, mode = 28 3. mean = 98.4, median = 103, mode = 99 4. mean = 5.4, median = 5.6, mode = 2.3 5. mean = 584.6, median = 594, mode = none 6. C

© Glencoe/McGraw-Hill   40   Glencoe Pre-Algebra

NAME _____ DATE _____ PERIOD _____

5-9 Solving Equations with Rational Numbers (Pages 244–248)

You can solve rational number equations using the same skills you used to solve equations involving integers.

Solving Equations	• Solving an equation means getting the variable alone on one side of the equation to find its value. • To get the variable alone, you use inverse operations to undo what has been done to the variable. • Addition and subtraction are inverse operations. • Multiplication and division are inverse operations. • Whatever you do to one side of the equation, you must also do to the other side to maintain the equality.

Examples

a. Solve $x + 5.7 = 2.5$.

$x + 5.7 = 2.5$
$x + 5.7 - 5.7 = 2.5 - 5.7$ Subtract 5.7 from each side.
$x = -3.2$ Simplify.

b. Solve $\frac{2}{3}y = \frac{5}{6}$.

$\frac{2}{3}y = \frac{5}{6}$
$\frac{3}{2}\left(\frac{2}{3}y\right) = \frac{3}{2}\left(\frac{5}{6}\right)$ Multiply each side by $\frac{3}{2}$.
$y = \frac{5}{4}$ or $1\frac{1}{4}$ Simplify.

Try These Together

1. Solve $\frac{3}{5} = a - \frac{1}{8}$.

   HINT: Add $\frac{1}{8}$ to each side.

2. Solve $1.4n = 4.2$.

   HINT: Divide each side by 1.4.

Practice

Solve each equation. Check your solution.

3. $p - 3.7 = -2.4$
4. $b - (-60.25) = 121.6$
5. $-8.8 + q = 14.3$
6. $w + \frac{1}{2} = \frac{7}{8}$
7. $j - \left(-\frac{1}{9}\right) = \frac{1}{6}$
8. $y - 1\frac{2}{5} = 2\frac{4}{5}$
9. $-5y = 8.5$
10. $-2.7t = -21.6$
11. $4.2d = -10.5$
12. $9z = \frac{3}{4}$
13. $\frac{m}{5} = -\frac{1}{10}$
14. $-\frac{5}{6}a = 20$

15. **Standardized Test Practice** Solve for the measure of x.
    - **A** 4.5 m
    - **B** 4.4 m
    - **C** 3.5 m
    - **D** 3.4 m

    (Rectangle: 25.2 m top, 21.7 m bottom with x segment)

Answers: 1. $\frac{29}{40}$ 2. 3 3. 1.3 4. 61.35 5. 23.1 6. $\frac{3}{8}$ 7. $\frac{1}{18}$ 8. $4\frac{1}{5}$ 9. -1.7 10. 8 11. -2.5 12. $\frac{1}{12}$ 13. $-\frac{1}{2}$ 14. -24 15. C

5-10 Arithmetic and Geometric Sequences (Pages 249–252)

A branch of mathematics called **discrete mathematics** deals with topics like logic and statistics. Another topic of discrete mathematics is **sequences**. A sequence is a list of numbers in a certain order. Each number is called a **term** of the sequence. When the difference between any two consecutive, or side-by-side, terms is the same, that difference is the **common difference** and the sequence is an **arithmetic sequence**.

A sequence of numbers such as 1, 2, 4, 8, 16, 32, 64 forms a **geometric sequence**. Each number in a geometric sequence increases or decreases by a common *factor* called the **common ratio**.

Examples

a. Is the sequence 4, 12, 36, 108, ... geometric? If so, state the common ratio and list the next two terms.

Notice that 4 × 3 = 12, 12 × 3 = 36, and 36 × 3 = 108.
4 12 36 108
 ×3 ×3 ×3

This sequence is geometric with a common ratio of 3. The next two terms are 108 × 3 or 324 and 324 × 3 or 972.

b. Is the sequence 2, 5, 8, 11, ... arithmetic? What are the next 3 terms?

2 5 8 11
5−2 8−5 11−8
or or or
+3 +3 +3

Since the difference between any two consecutive terms is the same, the sequence is arithmetic.

...8 11 14 17 20
 +3 +3 +3 +3

Continue the sequence to find the next three terms.

Practice

State whether each sequence is arithmetic or geometric. Then write the next three terms of each sequence.

1. −2, −4, −8, −16, ...
2. 10, 5, 0, −5, −10, ...
3. 35, 28, 21, 14, ...
4. 1, 3, 9, 27, ...   *geometric −32, −64, −128*
5. 0.5, 0.8, 1.1, 1.4, ...
6. −8, −6, −4, −2, ...
7. 0.5, 1.5, 4.5, 13.5, ...
8. 2, −4, 8, −16, ...
9. $\frac{1}{5}, \frac{2}{5}, \frac{3}{5}, \frac{4}{5}$ ...

10. **Standardized Test Practice** Find the next three terms in the sequence 8, 16, 24, 32, ...

A 32, 24, 16
B 40, 48, 56
C 44, 56, 64
D 64, 128, 264

Answers: 1. geometric; −32, −64, −128 2. arithmetic; −15, −20, −25 3. arithmetic; 7, 0, −7 4. geometric; 81, 243, 729 5. arithmetic; 1.7, 2, 2.3 6. arithmetic; 0, 2, 4 7. geometric; 40.5, 121.5, 364.5 8. geometric; 32, −64, 128 9. arithmetic; 1, $1\frac{1}{5}$, $1\frac{2}{5}$ 10. B

© Glencoe/McGraw-Hill   Glencoe Pre-Algebra

5 Chapter Review

Pizza Pig-out

1. A group of five friends ordered a pizza to share. Solve each equation to find out what portion of the pizza each person ate. Write your answer in the blank beside the person's name.

   _____ Andrew $a - 5.1 = -4.8$ +5.1 +5.1 $a = 0.3$

   _____ Nancy $n - (-10.95) = 11.1$

   _____ Jocelyn $\frac{6}{30} + j = \frac{18}{30}$

   _____ Samantha $s - \frac{1}{2} = -\frac{4}{10}$

   _____ Mark $m + 1\frac{1}{5} = 1\frac{1}{4}$

2. Change your answers for Andrew and Nancy to fractions and write the 5 fractions in order from least to greatest.

3. Who ate the most and who ate the least?

4. Draw a pizza and divide it into 5 slices showing how much each person ate. Use your list in Exercise 2 to help estimate the sizes of each slice. Label each slice with the person's name and the amount they ate.

Answers are located in the Answer Key.

© Glencoe/McGraw-Hill   Glencoe Pre-Algebra

6-1 Ratios and Rates (Pages 264–268)

A **ratio** is a comparison of two numbers by division. The ratio of the number 2 to the number 3 can be written in these ways: 2 to 3, 2:3, or $\frac{2}{3}$. Ratios are often expressed as fractions in simplest form or as decimals.

| Rates | A **rate** is a special ratio that compares two measurements with different units of measure, such as miles per gallon or cents per pound. A rate with a denominator of 1 is called a **unit rate**. |

Example
Jane buys 6 cans of soda for $1.74. Express this as a unit rate for 1 soda.

First write the ratio as a fraction: $\frac{\$1.74}{6\text{ sodas}}$. Then divide the numerator and denominator by 6.

$\frac{\$1.74}{6 \text{ cans}} = \frac{\$0.29}{1 \text{ can}}$ (÷6)   Thus, one can of soda costs $0.29.

Try These Together
1. Express the ratio 2 to 28 as a fraction in simplest form.
2. Express the ratio $210 for 5 nights as a unit rate.

Practice
Express each ratio or rate as a fraction in the simplest form.

3. 10:35
4. 60:20
5. 3 to 39
6. 8 out of 14
7. 18 boys to 15 girls
8. 16 blue to 4 green

Express each ratio as a unit rate.

9. 294 miles on 10 gallons
10. $0.72 for 12 ounces
11. $3.88 for 2 pounds
12. 3.4 inches of rain in 2 months
13. 200 meters in 23.5 seconds
14. $21 for a half dozen roses
15. $60 for 8 movie tickets
16. 6 limes for $2

17. **Consumer Awareness** You are trying to decide whether to buy a package of 20 yellow pencils for $1.25 or a package of 15 rainbow pencils for $1.09. Which one is a better buy and why?

18. **Standardized Test Practice** The temperature increased 12°F in 48 hours. How can the temperature increase be described with a unit rate?

A $\frac{10°F}{40 \text{ h}}$ B $\frac{1°F}{4 \text{ h}}$ C $\frac{0.25°F}{\text{h}}$ D $\frac{1°F}{0.25 \text{ h}}$

Answers: 1. $\frac{1}{14}$ 2. $42/night 3. $\frac{2}{7}$ 4. 3 5. $\frac{1}{13}$ 6. $\frac{4}{7}$ 7. $\frac{6}{5}$ 8. 4 9. 29.4 mi/gal 10. $0.06/oz 11. $1.94/lb 12. 1.7 in./mo 13. about 8.5 m/sec 14. $3.50 per rose 15. $7.50 per ticket 16. 3 limes per dollar 17. 20 yellow pencils because they cost about $0.06 each and the 15 rainbow pencils cost about $.07 each 18. C

NAME _____ DATE _____ PERIOD _____

6-2 Using Proportions (Pages 270–274)

A **proportion** is a statement that two or more ratios are equal, as in $\frac{a}{b} = \frac{c}{d}$.

The products ad and cb are called the **cross products** of the proportion. One way to determine if two ratios form a proportion is to check their cross products.

Property of Proportions	The cross products of a proportion are equal. If $\frac{a}{b} = \frac{c}{d}$, then $ad = cb$. If $ad = cb$, then $\frac{a}{b} = \frac{c}{d}$.

Examples

a. Solve $\frac{6}{y} = \frac{3}{2}$.

$6 \cdot 2 = y \cdot 3$ Cross products
$12 = 3y$ Multiply.
$\frac{12}{3} = \frac{3y}{3}$ Divide each side by 3.
$4 = y$

The solution is 4.

b. Replace the ● with = or ≠ to make a true statement.

$\frac{2}{5}$ ● $\frac{28}{70}$

Examine the cross products.

$2 \cdot 70 \stackrel{?}{=} 5 \cdot 28$
$140 = 140$

Since the cross products are equal, $\frac{2}{5} = \frac{28}{70}$.

Practice

Replace each ● with = or ≠ to make a true statement.

1. $\frac{2}{5}$ ● $\frac{8}{20}$ 2. $\frac{3}{4}$ ● $\frac{18}{24}$ 3. $\frac{2.5}{7.5}$ ● $\frac{2}{6}$ 4. $\frac{84}{96}$ ● $\frac{7}{8}$ 5. $\frac{1}{5}$ ● $\frac{19}{90}$

Solve each proportion.

6. $\frac{x}{5} = \frac{77}{35}$ 7. $\frac{6}{m} = \frac{1}{36}$ 8. $\frac{12}{17} = \frac{n}{68}$ 9. $\frac{45}{x} = \frac{2}{3}$ 10. $\frac{4}{7} = \frac{5.2}{x}$

Write a proportion that could be used to solve for each variable. Then solve the proportion.

11. 3 pounds for $2.50
    2 pounds for n dollars

12. 3 notepads have 144 sheets
    x notepads have 240 sheets

13. **Food** To make a fruit salad, Jeff will use 3 oranges for every 2 people. If the salad is to serve 12 people, how many oranges will he use?

14. **Standardized Test Practice** A display case of old CDs are marked 2 for $15. If you pick out 5 CDs, how much will they cost, not including tax?
    A $67.50   B $60   C $38   D $37.50

NAME _____ DATE _____ PERIOD _____

6-3 Scale Drawings and Models *(Pages 276–280)*

When objects are too small or too large to be drawn or constructed at actual size, people use a **scale drawing** or a **model**. The **scale** is the relationship between the measurements of the drawing or model to the measurements of the object. The scale can be written as a **scale factor**, which is the ratio of the length or size of the drawing or model to the length of the corresponding side or part on the actual object.

Examples

a. The key on a map states that 1 inch is equal to 10 miles. Write the scale for the map.

$\frac{1 \text{ inch}}{10 \text{ miles}}$ Write a fraction as $\frac{\text{inches}}{\text{miles}}$.

b. According to EXAMPLE A, how far apart would two cities be in actual distance if they were 5 inches apart on the map?

$\frac{1}{10} = \frac{5}{x}$ Write a proportion using the scale.
$1 \cdot x = 5 \cdot 10$ Use the property of proportions.
$x = 50 \text{ mi}$ Solve.

Practice

On a set of blueprints for a new home, the contractor has established a scale that states $\frac{1}{2}$ inch = 10 feet. Use this information for problems 1–6.

1. What is the actual length of the living room whose distance is 1 inch on the blueprints?

2. What is the actual width of the living room whose distance is $\frac{3}{4}$ inch on the blueprints?

3. What is the actual height of the living room whose distance is $\frac{9}{20}$ inch on the blueprints?

4. If the buyer would like a kitchen to be 18 feet in length, how long should the kitchen be in the blueprints?

5. What are the dimensions on the blueprints of a bedroom that will be 18 feet by 16 feet when the house is built?

6. If the den has dimensions of 0.5 inch by 0.6 inch on the blueprints, what will be the dimensions of the actual den after the house is built?

7. **Standardized Test Practice** A model car has a scale of 1:24, where the model dimensions are in the numerator and the actual car dimensions are in the denominator. If the tires on the model have a diameter of $\frac{1}{2}$ inch, how long is the diameter of an actual tire on the car?

  A 9 inches   B 10 inches   C 12 inches   D 20 inches

Answers: 1. 20 feet 2. 15 feet 3. 9 feet 4. 0.9 inch 5. 0.9 inch by 0.8 inch 6. 10 feet by 12 feet 7. C

© Glencoe/McGraw-Hill   46   Glencoe Pre-Algebra

6-4 Fractions, Decimals, and Percents

(Pages 281–285)

| Writing Equivalent Forms of Fractions, Decimals, and Percents | • To express a decimal as a percent, write the decimal as a fraction with 1 as the denominator. Then write that fraction as an equivalent fraction with 100 as the denominator.
• To express a fraction as a percent, first write the fraction as a decimal by dividing numerator by denominator. Then write the decimal as a percent.
• To express a percent as a fraction, write the percent in the form $\frac{r}{100}$ and simplify. To express a percent as a decimal, write the percent in the form $\frac{r}{100}$ and then write as a decimal. |

Examples

a. Express $\frac{3}{5}$ as a decimal and as a percent.

$\frac{3}{5} = 3 \div 5$ $0.6 = 0.60$

$\phantom{\frac{3}{5}} = 0.6$ $ = \frac{60}{100}$ or 60%

b. Express 0.08 as a fraction and as a percent.

$\frac{0.08}{1} \xrightarrow{\times 100} \frac{8}{100}$ or $\frac{2}{25}$ $0.08 = \frac{8}{100}$

$\phantom{\frac{0.08}{1}} \xleftarrow{\times 100}$ $ = 8\%$

Try These Together

1. Express 0.59 as a percent and then as a fraction.
   HINT: Begin by writing 0.59 as $\frac{59}{100}$.

2. Express 45% as a decimal and then as a fraction.
   HINT: 45% means how many out of 100?

Practice

Express each decimal as a percent and then as a fraction in simplest form.

3. 0.90   4. 0.80   5. 1.35   6. 3.20   7. 0.62   8. 2.24

Express each fraction as a percent and then as a decimal.

9. $\frac{3}{6}$ 10. $\frac{2}{5}$ 11. $\frac{12}{16}$ 12. $\frac{5}{4}$ 13. $\frac{18}{40}$ 14. $\frac{1}{8}$

15. **Retail** A floor lamp is on sale for 60% off. What fraction off is this?

16. **Standardized Test Practice** Which of the following lists is in order from least to greatest?
    A 2.5, 2.5%, 0.0025
    B 2.5%, 0.25, 2.5
    C 0.0025, 0.25, 2.5%
    D 0.25, 2.5%, 2.5

Answers: 1. 59%; $\frac{59}{100}$ 2. 0.45; $\frac{9}{20}$ 3. 90%; $\frac{9}{10}$ 4. 80%; $\frac{4}{5}$ 5. 135%; $\frac{27}{20}$ 6. 320%; $\frac{16}{5}$ 7. 62%; $\frac{31}{50}$ 8. 224%; $\frac{56}{25}$ 9. 50%; 0.5   10. 40%; 0.4   11. 75%; 0.75   12. 125%; 1.25   13. 45%; 0.45   14. 12.5%; 0.125   15. $\frac{3}{5}$ 16. B

NAME _____ DATE _____ PERIOD _____

6-5 Using the Percent Proportion (Pages 288–292)

A percent is a ratio that compares a number to 100. Percent also means *hundredths*, or *per hundred*. The symbol for percent is %.

The Percent Proportion	The **percent proportion** is $\frac{part}{base} = \frac{percent}{100}$. In symbols $\frac{a}{b} = \frac{p}{100}$, where *a* is the part, *b* is the base, and *p* is the percent.

Examples

a. Express $\frac{2}{5}$ as a percent.

$\frac{a}{b} = \frac{p}{100} \rightarrow \frac{2}{5} = \frac{p}{100}$ Replace a with 2 and b with 5.
$2 \cdot 100 = 5 \cdot p$ Find the cross products.
$\frac{200}{5} = \frac{5p}{5}$ Divide each side by 5.
$40 = p$

$\frac{2}{5}$ is equivalent to 40%.

b. 13 is 26% of what number?

$\frac{a}{b} = \frac{p}{100} \rightarrow \frac{13}{b} = \frac{26}{100}$ Replace a with 2 and p with 26.
$13 \cdot 100 = b \cdot 26$ Find the cross products.
$\frac{1300}{26} = \frac{26b}{26}$ Divide each side by 26.
$50 = b$

13 is 26% of 50.

Try These Together

1. Express $\frac{5}{8}$ as a percent.

   HINT: Use the proportion $\frac{5}{8} = \frac{p}{100}$.

2. What is 30% of 20?

   HINT: The value after the word "of" is usually the base.

Practice

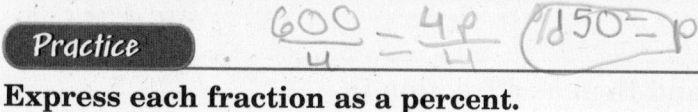

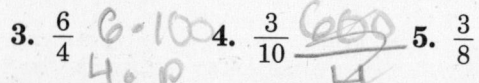

Express each fraction as a percent.

3. $\frac{6}{4}$ 4. $\frac{3}{10}$ 5. $\frac{3}{8}$ 6. $\frac{4}{25}$ 7. $\frac{17}{20}$ 8. $\frac{8}{5}$

Use the percent proportion to solve each problem.

9. 14 is what percent of 50?
10. 27 is what percent of 90?
11. 120 is what percent of 200?
12. 14 is 20% of what number?
13. 17 is 8.5% of what number?
14. 43 is 10% of what number?
15. What is 8% of 75?
16. What is 300% of 12?
17. Find 51% of $80.
18. Find 30% of $10.69.

19. **Retail** A pair of $32 jeans is marked down 40%. What is 40% of $32? What is the price of the jeans after the reduction?

20. **Standardized Test Practice** Bonnie got 12 out of 16 questions correct on her math quiz. What percent did she get correct?

   A $133\frac{1}{3}$%   B 75%   C 60%   D 25%

Answers: 1. 62.5% 2. 6 3. 150% 4. 30% 5. 37.5% 6. 16% 7. 85% 8. 160% 9. 28% 10. 30% 11. 60% 12. 70 13. 200 14. 430 15. 6 16. 36 17. $40.80 18. $3.21 19. $12.80; $19.20 20. B

© Glencoe/McGraw-Hill   48   Glencoe Pre-Algebra

6-6 Finding Percents Mentally (Pages 293–297)

When an exact answer is not needed, you can estimate percentages.

Estimating Percents	Method 1: With the fraction method, use a fraction that is close to the percent. For example, 24% is about 25% or $\frac{1}{4}$.
	Method 2: With the 1% method, find 1% of the number. Round the result, if necessary, and then multiply to find the percentage.
	Method 3: Use the meaning of percent to estimate.

Examples

a. Estimate 40% of 183 using the 1% method.
1% of 183 is 1.83 or about 2.
So 40% of 183 is about 40 × 2 or 80.

b. Estimate 60% of 537 using the meaning of percent.
60% means 60 for every 100 or 6 for every 10.
537 has 5 hundreds and about 4 tens (37 ≈ 40).
(60 × 5) + (6 × 4) = 300 + 24 or 324.

Try These Together

1. What fraction could you use to estimate 34% of a number?
   HINT: $\frac{1}{3}$ is about 33%.

2. Estimate a percent for 29 out of 40.
   HINT: 29 out of 40 is close to 30 out of 40.

Practice

Write the fraction, mixed number, or whole number you could use to estimate.

3. 110%   4. 22%   5. 41%
6. 8.5%   7. 49%   8. 430%

Estimate.

9. 13% of 79   10. 58% of 190   11. 98% of 11   12. 41% of 20
13. 109% of 500   14. 73% of 21   15. 87% of 90   16. 31% of 87

Estimate each percent.

17. 19 out of 39   18. 20 out of 55   19. 4 out of 300

20. **Nutrition** If a package of 4 cookies has 205 Calories and 30% of the Calories come from fat, estimate how many of the 205 Calories are from fat.

21. **Standardized Test Practice** Choose the best estimate for 11% of 833.
   A 0.083   B 0.83   C 8.3   D 83

Answers: Estimates may vary. 1. $\frac{1}{3}$ 2. 75% 3. $1\frac{1}{10}$ 4. $\frac{1}{5}$ 5. $\frac{2}{5}$ 6. $\frac{1}{10}$ 7. $\frac{1}{2}$ 8. $4\frac{3}{10}$ 9. 8 10. 120 11. 11 12. 8 13. 550 14. 15 15. 81 16. 27 17. 50% 18. 40% or 33% 19. 1% 20. 60 calories 21. D

6-7 Using Percent Equations (Pages 298–302)

Interest (I) is money earned or paid for the use of an amount of money, called the **principal** (p), at a stated rate (r), or percent, for a given amount of time (t). Interest can be calculated using the formula $I = prt$. Another common use of percent is with a **discount**, or amount of money deducted from a price.

Percent Equation	The formula is Part = Percent · Base.

Examples

a. What is the discount if a $6.40 item is on sale for 30% off?

Write in Part = Percent · Base form.
What is 30% of $6.40?
Part = 0.30 × 6.40
Part = 1.92   The discount is $1.92.

b. Find the interest on $460 invested at 8% annually for 2 years.

$I = prt$ Interest formula
$I = (460)(0.08)(2)$ $p = 460, r = 8\%$ or $0.08, t = 2$
$I = 73.6$
The interest is $73.60.

Try These Together

1. Use Part = Percent · Base to find what percent 34 is of 80.
   HINT: 34 is the percent and 80 is the base.

2. What is the discount if a $45 item is on sale at 15% off?
   HINT: To find 15% of $45, multiply $45 by 0.15.

Practice

Solve each problem by using the percent equation, Part = Percent · Base.

3. 56 is what percent of 64?
4. 70 is 40% of what number?
5. 30 is 60% of what number?
6. What is 33% of 60?
7. What is 40% of 350?
8. Find 60% of $8.99.

Find the discount or interest to the nearest cent.

9. $3.99 socks, 40% off
10. $250 desk, 30% off
11. $15 wrist watch, 75% off
12. $20 telephone, 25% off
13. $1400 at 2% interest monthly for 30 months
14. $650 at 9% interest annually for 2 years

15. **Standardized Test Practice** After October 31, you find the holiday candy marked down 70%. How much money would you save if your favorite candy regularly costs $2.99?
    A $2.29   B $2.09   C $0.90   D $0.70

Answers: 1. 42.5% 2. $6.75 3. 87.5% 4. 175 5. 50 6. 19.8 7. 140 8. $5.39 9. $1.60 10. $75 11. $11.25 12. $5 13. $840 14. $117 15. B

NAME _____ DATE _____ PERIOD _____

6-8 Percent of Change *(Pages 304–308)*

The **percent of change** is the ratio of the amount of change to the original amount. When an amount increases, the percent of change is a **percent of increase**. When the amount decreases, the percent of change is negative. You can also state a negative percent of change as a **percent of decrease**.

Percent of Change	percent of change = $\dfrac{\text{amount of change}}{\text{original measurement}}$

Examples

a. What is the percent of change from 30 to 24?

amount of change = new − old
= 24 − 30 or −6

percent of change = $\dfrac{\text{amount of change}}{\text{original measurement}}$

= $\dfrac{-6}{30}$

= −0.2 or −20%

The percent of change is −20%.
The percent of decrease is 20%.

b. What is the percent of change from 8 to 10?

amount of change = new − old
= 10 − 8 or 2

percent of change = $\dfrac{\text{amount of change}}{\text{original measurement}}$

= $\dfrac{2}{8}$

= 0.25 or 25%

The percent of change is 25%.
The percent of increase is 25%.

Practice

State whether each percent of change is a percent of increase or a percent of decrease. Then find the percent of increase or decrease. Round to the nearest whole percent.

1. old: 2 rabbits
   new: 13 rabbits

2. old: 125 people
   new: 90 people

3. old: 10 minutes
   new: 25 minutes

4. old: 1000 widgets
   new: 540 widgets

5. old: $5,000
   new: $4,700

6. old: 140 pounds
   new: 155 pounds

7. old: 15 centimeters
   new: 17 centimeters

8. old: $32.99
   new: $23.09

9. old: $1250
   new: $1310

10. **Safety** If a manufacturer reduces the number of on-the-job accidents from an average of 20 a month to an average of 6 a month, what is the percent of decrease in accidents?

11. **Standardized Test Practice** If the price of gas increases from $1.01 per gallon to $1.21 per gallon, what is the percent of increase?
    A 19%   B 20%   C 21%   D 22%

Answers: 1. increase 550% 2. decrease 28% 3. increase 150% 4. decrease 46% 5. decrease 6% 6. increase 11% 7. increase 13% 8. decrease 30% 9. increase 5% 10. 70% decrease 11. B

© Glencoe/McGraw-Hill   51   Glencoe Pre-Algebra

NAME _____ DATE _____ PERIOD _____

6-9 Probability and Predictions (Pages 310–314)

Probability is the chance that some event will happen. It is the ratio of the number of ways an event can occur to the number of possible outcomes. The set of all possible outcomes is called the **sample space**.

Probability	Probability = $\dfrac{\text{number of ways a certain outcome can occur}}{\text{number of possible outcomes}}$ The probability of an event, P(event), is always between 0 and 1, inclusive.

Examples A bowl contains 7 slips of paper with the name of a day of the week on each slip.

a. If you draw a slip from the bowl, what is the probability that the day contains the letter "y"?

The probability of an event that is certain is 1. Since the name of every day of the week contains a "y," this probability is 1.

b. What is the probability that you draw a day of the week that contains the letter "s"?

Five days of the week have the letter "s." The probability of drawing a day with this letter is $\dfrac{5}{7}$.

Try These Together

1. What is the probability that a 5 is rolled on a number cube?
   HINT: A 5 can occur in only 1 way on a single number cube. There are 6 possible outcomes.

2. Find the probability that a number greater than 6 is rolled on a number cube.
   HINT: When an event is certain not to happen, the probability is 0.

Practice

Suppose the numbers from 1 to 20 are written on 20 slips of paper and put into a bowl. You draw a slip at random. State the probability of each outcome.

3. The number is less than 5. $\dfrac{1}{5}$
4. The number ends in 5.
5. The number is even.
6. The number is divisible by 3.
7. The number is prime.
8. The digits have a sum of 10.
9. The number is less than 25.
10. The number contains a "1."

There are 5 purple marbles, 7 gold marbles, and 3 red marbles in a bag. Suppose one marble is chosen at random. Find each probability.

11. P(gold)
12. P(purple)
13. P(red or gold)
14. P(not red)

15. **Standardized Test Practice** What is the probability of rolling a number other than a 1 or 2 on a number cube?

    A $\dfrac{5}{6}$ B $\dfrac{2}{3}$ C $\dfrac{1}{2}$ D $\dfrac{1}{3}$

Answers: 1. $\dfrac{1}{6}$ 2. 0 3. $\dfrac{1}{5}$ 4. $\dfrac{1}{10}$ 5. $\dfrac{1}{2}$ 6. $\dfrac{3}{10}$ 7. $\dfrac{2}{5}$ 8. $\dfrac{1}{20}$ 9. 1 10. $\dfrac{11}{20}$ 11. $\dfrac{7}{15}$ 12. $\dfrac{1}{3}$ 13. $\dfrac{2}{3}$ 14. $\dfrac{4}{5}$ 15. B

© Glencoe/McGraw-Hill   Glencoe Pre-Algebra

NAME _____ DATE _____ PERIOD _____

6 Chapter Review
Mad Lib Math

You and your parent or guardian can play a game of mad lib math. Your parent will ask you for the information requested in parentheses and fill in each blank in the paragraph below. Then read the paragraph and then answer the questions that follow.

_____ and _____ out of _____ of his/her friends went to
1. (your name)   2. (ratio)

a carnival one afternoon. _____ tried the Test of Strength
                           3. (name a friend)

and could only get the bell ringer to raise _____ feet high.
                                            4. (decimal greater than 1)

_____ spent _____ trying to win a teddy
5. (name a friend)   6. (dollars and cents)

bear. At the dunking booth _____ dunked the heckler
                           7. (name a friend)

_____ out of _____ times. _____ and his/her best friend
8. (ratio)   9. (your name)

raced against each other and _____ won by a margin of
                             10. (your name)

_____ second. By the end of the afternoon, they had all spent
11. (decimal less than 1)

_____ of their money and they decided it was time to go
12. (percent less than 100%)

home.

13. Express the ratio in Exercise 8 as a decimal.

14. Express the ratio in Exercise 2 as a percent.

15. Express the percent in Exercise 12 as a fraction.

16. If 2 drinks at the carnival cost $3.20, how much will 9 drinks cost?

17. If 1700 people attended the carnival that day, and 3 out of 5 of them were male, how many of the attendees that day were male?

18. Suppose you took $20 with you to the carnival and came home with $3.50.
    a. $3.50 is what percent of $20?
    b. Find the percent of decrease.

Answers are located in the Answer Key.

© Glencoe/McGraw-Hill   53   Glencoe Pre-Algebra

NAME _____ DATE _____ PERIOD _____

7-1 Solving Equations with Variables on Each Side *(Pages 330–333)*

Sometimes we encounter equations that have a variable on both sides of the equal sign. If this situation occurs, then use the Addition or Subtraction Property of Equality to rewrite the equation with a variable on only one side. Once the equation is written with one variable, it can be solved using inverse operations.

Examples

a. Solve $4x + 1 = 2x + 5$.

$4x + 1 - 2x = 2x + 5 - 2x$	Subtraction Property
$2x + 1 = 5$	Simplify.
$2x + 1 - 1 = 5 - 1$	Subtraction Property
$2x = 4$	Simplify.
$\frac{2x}{2} = \frac{4}{2}$	Division Property
$x = 2$	Simplify to solve.

b. Solve $q - 3 = -3q - 43$.

$q - 3 - q = -3q - 43 - q$
$-3 = -4q - 43$
$-3 + 43 = -4q - 43 + 43$
$40 = -4q$
$\frac{40}{-4} = \frac{-4q}{-4}$
$-10 = q$

Practice

Solve each equation.

1. $5x + 1 = 4x - 1$
2. $-10b + 5 = 7b + 5$
3. $r + 15 = 4r - 6$
4. $10 - 2v = -5v - 50$
5. $15y + 3 = 18y$
6. $-2x + 6 = 4x + 9$

Write an equation then solve.

7. Four more than -3 times a number is equal to 8 more than -4 times the same number.

8. Twice a number decreased by one equals the same number added to two.

9. Six plus -2 times a number is the same as 26 plus six times the same number.

10. Negative ten times a number minus five equals negative eleven times the same number.

11. **Standardized Test Practice** Solve the equation $4x + 3 = -2x - 99$ for the variable x.
    A -17 B 17 C -48 D 48

Answers: 1. -2 2. 0 3. 7 4. -20 5. 1 6. $-\frac{1}{2}$ 7. 4 8. 3 9. $-\frac{5}{2}$ 10. 5 11. A

© Glencoe/McGraw-Hill   54   Glencoe Pre-Algebra

NAME _____ DATE _____ PERIOD _____

7-2 Solving Equations with Grouping Symbols (Pages 334–338)

Some equations have the variable on each side of the equals sign. Use the properties of equality to eliminate the variable from one side. Then solve the equation. You may find that some equations have *no solution*. The solution set is the **null** or **empty set**. It is shown by the symbol { } or ∅.

Examples Solve each equation.

a. $12 + 3a = 7a$

$12 + 3a - 3a = 7a - 3a$ Subtract 3a from each side.

$\frac{12}{4} = \frac{4a}{4}$ Divide each side by 4.

$3 = a$ The solution is 3.

b. $4b - 7 = 13 + 4b$

$4b - 4b - 7 = 13 + 4b - 4b$ Subtract 4b from each side.

$-7 = 13$

This sentence is never true, so there is no solution for this equation. The solution set is ∅.

Try These Together
Solve each equation.

1. $5t = 3 + t$ **2.** $6g - 4 = g + 1$ **3.** $c = 4c + 8$

HINT: Eliminate the variable from one side of the equation then solve.

Practice
Solve each equation.

4. $9h - 3 = h$

5. $-16d + 4 = d$

6. $7m = 18m - 2$

7. $6 + 3(1 + 3a) = 2a$

8. $n + 8 = -5 + 4n$

9. $4 - 2(2 + 4x) = x - 3$

10. $8p - 2p + 3 = 10p - 6$

11. $15 + 5(w - 2) = 7w + 4$

12. $12r + 34 = -6r - (-9)$

13. $6k + 3(k + 2) = 5k + 12$

14. $2s - 4.2 = -8s + 8$

15. $7x + \frac{1}{-8} = x - \frac{3}{4}$

16. Geometry Find the dimensions of the rectangle if the perimeter is 118 feet.

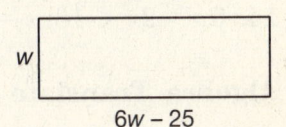

     [rectangle with width w and length $6w - 25$]

17. Algebra Eight times a number plus two is five times the number decreased by three. What is the number?

18. Standardized Test Practice Solve the equation $4k + 2(k + 1) = 3k + 4$.

   A $\frac{2}{3}$ **B** 2   **C** 4   **D** 6

Answers: 1. $\frac{3}{4}$ 2. 1 3. $-2\frac{2}{3}$ 4. $\frac{3}{8}$ 5. $\frac{4}{17}$ 6. $\frac{2}{11}$ 7. $-1\frac{7}{2}$ 8. $4\frac{1}{3}$ 9. $\frac{1}{3}$ 10. $2\frac{1}{4}$ 11. $\frac{1}{2}$ 12. $-1\frac{7}{18}$ 13. $1\frac{1}{2}$ 14. 1.22 15. $-\frac{5}{48}$ 16. width: 12 ft, length: 47 ft 17. $-1\frac{2}{3}$ 18. A

© Glencoe/McGraw-Hill   Glencoe Pre-Algebra

7-3 Inequalities (Pages 340–344)

A mathematical sentence that contains $<$, $>$, $\leq$, or $\geq$ is called an **inequality**. Inequalities, like equations, can be true, false, or open. Most situations in real life can be described using inequalities. The table below shows some common phrases and corresponding inequalities.

$<$	$>$	$\leq$	$\geq$
• less than • fewer than	• greater than • more than • exceeds	• less than or equal to • no more than • at most	• greater than or equal to • no less than • at least

Examples

a. State whether $2y < 12$ is true, false, or open.

$2y < 12$

Until the variable y is replaced by a number, this inequality is open.

b. Translate the sentence "5 times a number is greater than or equal to 75," into an inequality.

Let n represent the number. Then translate the words into an inequality using the variable.

five times number is greater than or equal to 75
 5   ×   n   ≥   75

$5n \geq 75$

Practice

State whether each inequality is *true, false,* or *open*.

1. $3 > 7$
2. $y \leq 8$
3. $1 \geq 1$
4. $2n > 18$
5. $12 > 10$
6. $1 < 4x$
7. $8 > 16$
8. $6 \leq 8$
9. $2x > 7$
10. $32 < 40$

State whether each inequality is *true* or *false* for the given value.

11. $18 + z < 23$; $z = 8$
12. $m - 8 > 17$; $m = 29$
13. $3x < 14$; $x = 5$
14. $6x - 2x < 18$; $x = 3$
15. $18 \geq 6m$; $m = 3$
16. $j + 13 > 27$; $j = 7$

Algebra Translate each sentence into an inequality.

17. At least 18 people were at the party.
18. There were less than 5 A's.
19. The crowd was made up of more than 80 people.

20. **Standardized Test Practice** The Super Bowl is the most viewed sports event televised every year. There are over one billion viewers every year. Write an inequality to describe this situation.

   A $x > 1,000,000,000$
   B $x < 1,000,000,000$
   C $x = 1,000,000,000$
   D $x \leq 1,000,000,000$

Answers: 1. F 2. O 3. T 4. O 5. T 6. O 7. F 8. T 9. O 10. T 11. F 12. T 13. F 14. T 15. T 16. F 17. $x \geq 18$ 18. $x < 5$ 19. $x > 80$ 20. A

7-4 Solving Inequalities by Adding or Subtracting (Pages 345–349)

Solving inequalities that involve addition or subtraction is just like solving equations that involve addition or subtraction.

Addition and Subtraction Properties of Inequalities	Adding or subtracting the same number from each side of an inequality does not change the truth of the inequality. For all numbers a, b, and c: 1. If $a > b$, then $a + c > b + c$ and $a - c > b - c$. 2. If $a < b$, then $a + c < b + c$ and $a - c < b - c$. The rules for $a \geq b$ and $a \leq b$ are similar.

Examples

a. Solve $b + 18 > 53$.

$b + 18 > 53$
$b + 18 - 18 > 53 - 18$ Subtract 18 from each side.
$b > 35$ Check your solution by replacing b with a number greater than 35 in the original the inequality.

b. Solve $n - 32 \leq 6$.

$n - 32 \leq 6$
$n - 32 + 32 \leq 6 + 32$ Add 32 to each side.
$n \leq 38$ Check your solution by replacing n with 38 and a number less than 38 in original inequality.

Try These Together

Solve each inequality and check your solution.

1. $12 < n - 8$
2. $p - 9 \leq 14$
3. $c + (-8) > 2$

HINT: Adding the same number to each side or subtracting the same number from each side of an inequality does not change the truth of the inequality.

Practice

Solve each inequality and check your solution.

4. $t - (-7) \leq 21$
5. $33 \geq 13 + s$
6. $-19 < m - (-7)$
7. $46 \geq a + 14$
8. $r + (-5) > 27$
9. $k + 34 \geq 15$
10. $y - (-12) > 8$
11. $20 \leq x + 3$
12. $14 < z + (-8)$

13. **Driving** To pass the driver's test, you must complete both a written exam and a driving test. Your total score must be 70 or greater. Each portion of the test is worth 50 points. If you get a score of 40 on the written exam, what is the minimum score you must receive on the driving portion to pass the test?

14. **Standardized Test Practice** Tomás and Jan have saved $15,000 to buy a house. They have found a house they like that sells for $129,000. What is the least amount of money Tomás and Jan must borrow to buy the house?

A $144,000   B $114,000   C $100,000   D $500

Answers: 1. $20 < n$ 2. $p \leq 23$ 3. $c > 10$ 4. $t \leq 14$ 5. $s \leq 20$ 6. $-26 < m$ 7. $32 \geq a$ 8. $r > 32$ 9. $k \geq -19$ 10. $y > -4$ 11. $17 \leq x$ 12. $22 < z$ 13. 30 14. B

7-5 Solving Inequalities by Multiplying or Dividing (Pages 350–354)

Solving inequalities that involve multiplication or division is very similar to solving equations that involve multiplication or division. However, there is one very important difference involved with multiplying or dividing by negative integers.

Multiplication and Division Properties of Inequalities	When you multiply or divide each side of a true inequality by a *positive* integer, the result remains true. For all integers a, b, and c, where $c > 0$, if $a > b$, then $a \cdot c > b \cdot c$ and $\frac{a}{c} > \frac{b}{c}$.
Multiplication and Division Properties of Inequalities	When you multiply or divide each side of a true inequality by a *negative* integer, you must *reverse* the order symbol. For all integers a, b, and c, where $c < 0$, if $a > b$, then $a \cdot c < b \cdot c$ and $\frac{a}{c} < \frac{b}{c}$.

Examples

a. Solve $\frac{n}{7} < -7$.

$\frac{n}{7} < -7$

$\frac{n}{7} \cdot 7 < -7 \cdot 7$ Multiply each side by 7.

$n < -49$ Check your solution by replacing n with -56, a number less than -49.

b. Solve $-5m \geq 45$.

$-5m \geq 45$

$\frac{-5m}{-5} \leq \frac{45}{-5}$ Divide each side by -5 and reverse the order symbol.

$m \leq -9$ Check your solution by replacing m with -9 and a number less than -9.

Practice

Solve each inequality and check your solution.

1. $-3x \geq -24$
2. $6s \geq 30$
3. $\frac{x}{5} < 39$
4. $-162 < 18r$

5. $92 \geq -4p$
6. $-7y \geq 119$
7. $\frac{x}{3} > 16$
8. $\frac{b}{8} < 9$

9. $-6n \geq -72$
10. $15j \leq 135$
11. $18d < 126$
12. $8x \geq -72$

13. $4x \geq 36$
14. $\frac{y}{12} \leq 2$
15. $\frac{c}{8} > 2$
16. $-114 \leq -19r$

17. $\frac{m}{12} > 5$
18. $7 < \frac{n}{3}$
19. $-80 \leq -20s$
20. $38 \geq 19t$

21. **Standardized Test Practice** Dana will leave home at 9 A.M. and will drive to Titusville, which is 220 miles away. What is the least speed he must average to be sure he arrives in Titusville no later than 1 P.M.?

A 60 mph   B 55 mph   C 50 mph   D 45 mph

Answers: 1. $x \leq 8$ 2. $s \geq 5$ 3. $x < 195$ 4. $r > -9$ 5. $p \geq -23$ 6. $y \leq -17$ 7. $x > 48$ 8. $b < 72$ 9. $n \leq 12$ 10. $j \leq 9$ 11. $d < 7$ 12. $x \geq -9$ 13. $x \geq 9$ 14. $y \leq 24$ 15. $c > 16$ 16. $r \leq 6$ 17. $m > 60$ 18. $n > 21$ 19. $s \geq 4$ 20. $t \leq 2$ 21. B

7-6 Solving Multi-Step Inequalities (Pages 355–359)

You solve inequalities by applying the same methods you use to solve equations. Remember that if you multiply or divide each side of an inequality by a negative number, you must reverse the inequality symbol. When you solve inequalities that contain grouping symbols, you may need to use the distributive property to remove the grouping symbols.

Examples Solve each inequality.

a. $5y - 17 \leq 13$

$5y - 17 + 17 \leq 13 + 17$ Add 17 to each side.
$\frac{5y}{5} \leq \frac{30}{5}$ Divide each side by 5.
$y \leq 6$

b. $3(-5 - 2s) > 3$

$3(-5 - 2s) > 3$
$-15 - 6s > 3$ Use the distributive property.
$-15 - 6s + 15 > 3 + 15$ Add 15 to each side.
$\frac{-6s}{-6} < \frac{18}{-6}$ Divide each side by -6.
$s < 3$ Don't forget to reverse the inequality sign.

Try These Together
Solve each inequality.

1. $3x + 6 > 24$
2. $4x - 3 < 15$
3. $18 \leq 22 - 2n$

Practice
Solve each inequality.

4. $3x - 5 < 4x - 8$
5. $5b + 2 > 3b - 1$
6. $6k - 2 < 5k - 5$
7. $2.7g + 12 > 3.2g$
8. $6.9y - 2.2 < 3.9y - 1.3$
9. $18 + \frac{x}{5} \leq 20$
10. $16 - \frac{z}{6} \geq 24$
11. $\frac{c+5}{4} < \frac{10-c}{9}$
12. $\frac{n-7}{3} \leq -12$
13. $9a - (a + 2) > a + 17$
14. $\frac{b+2}{3} > \frac{b+4}{6}$
15. $\frac{6x+4}{3} > \frac{2x+7}{6}$

16. **Consumer Awareness** Ericel has $50 to spend for food for a birthday party. The birthday cake will cost $17, and he also wants to buy 4 bags of mixed nuts. Use the inequality $4n + 17 \leq 50$ to find how much he can spend on each bag of nuts.

17. **Standardized Test Practice** Solve the inequality $\frac{2x+4}{3} \leq \frac{3x+1}{5}$.

   A $x \leq 20$ **B** $x \leq 3$ **C** $x \leq -15$ **D** $x \leq -17$

Answers: 1. $x > 6$ 2. $x < 4\frac{1}{2}$ 3. $2 \geq n$ 4. $x > 3$ 5. $b > -1\frac{1}{2}$ 6. $k < -3$ 7. $g > 24$ 8. $y < 0.3$ 9. $x \leq 10$ 10. $z \leq -48$ 11. $c < -\frac{5}{13}$ 12. $n \leq -29$ 13. $a > 2\frac{5}{7}$ 14. $b > 0$ 15. $x > -\frac{1}{10}$ 16. at most $8.25 17. D

NAME _____ **DATE** _____ **PERIOD** _____

7 Chapter Review

Find the Hidden Picture

Solve each equation or inequality. Look for the solution in the solution code box at the bottom of the page. Then shade the sections of the picture that correspond with the correct solutions to the problems.

1. $x + 7 < 6$
2. $x - (-3) = -5$
3. $-24 = 6x$
4. $\dfrac{x}{-8} > -2$
5. $x - 6 > 10$
6. $2x - 5 = 7$
7. $-3x = 81$

For each listed value that is a solution to one of the equations above, shade in the corresponding section on the puzzle. For example, if $x = 30$ is a solution to one of the equations, shade in section 1 of the puzzle.

Value	Section	Value	Section
$x = 30$	1	$x > 13$	18
$x < 5$	2	$x \leq -17$	19
$x > -3$	3	$x = 2$	20
$x = 27$	4	$x = 11$	21
$x < -1$	5	$x < 8$	22
$x > 12$	6	$x > -4$	23
$x < 3$	7	$x > 5$	24
$x \leq 13$	8	$x = 19$	25
$x = -8$	9	$x \leq -22$	26
$x > -15$	10	$x = 6$	27
$x \leq 7$	11	$x < -11$	28
$x = -20$	12	$x = -27$	29
$x = 7$	13	$x > -9$	30
$x = -4$	14	$x = 31$	31
$x < 16$	15	$x < -14$	32
$x > 16$	16	$x \leq 23$	33
$x < 11$	17	$x > -7$	34

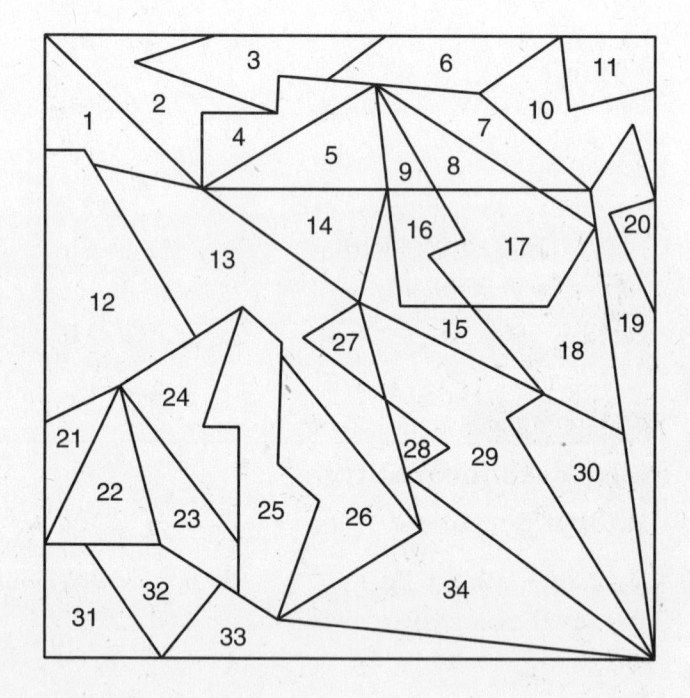

Answers are located in the Answer Key.

© Glencoe/McGraw-Hill   Glencoe Pre-Algebra

NAME _____ DATE _____ PERIOD ____

8-1 Functions (Pages 369–373)

A **relation** is a set of ordered pairs. The set of the first coordinates is the **domain** of the relation. The set of second coordinates is the **range** of the relation. You can model a relation with a table or graph.

Definition of a Function	A **function** is a relation in which each element in the domain is paired with exactly one element in the range. You can use the **vertical line test** to test whether a relation is a function.

Example

What are the domain and range of the relation graphed at the right? Is the relation a function?

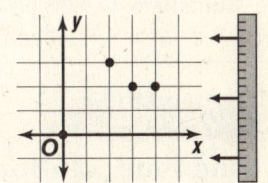

The set of ordered pairs for the relation is {(2, 3), (3, 2), (4, 2), (0, 0)}.
The domain is {2, 3, 4, 0}.
The range is {3, 2, 0}.
Since no vertical line passes through more than one point on the graph for any x-value, the relation is a function.

Try These Together

1. What is the domain and range of this relation? Is this relation a function? {(2, 7), (3, 8), (2, 1)}
   HINT: Is any x-value paired with more than one y-value?

2. What is the domain and range of this relation? Is this relation a function? {(11, −4), (−5, −3), (13, −3)}
   HINT: Is any x-value paired with more than one y-value?

Practice

Express the relation shown in each table or graph as a set of ordered pairs. State the domain and range of the relation. Then determine whether the relation is a function.

3.
x	y
0	3
−1	4
−3	9

4. [graph]

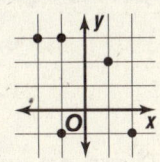

5. [graph]

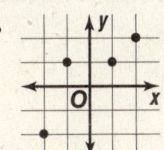

6. [graph]

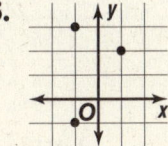

7. **Standardized Test Practice** What is the range of the relation {(7, 9), (10, 12)}?

   A {7, 10}   B {9, 12}   C {7, 12}   D {9, 10}

Answers: 1. D = {2, 3}, R = {1, 7, 8}, no 2. D = {−5, 11, 13}, R = {−4, −3}, yes 3. {(0, 3), (−1, 4), (−3, 9)}, D = {−3, −1, 0}, R = {3, 4, 9}, yes 4. {(−2, 3), (−1, 3), (−1, −1), (1, 2), (2, −1)}, D = {−2, −1, 1, 2}, R = {−2, −1, 2, 3}, no 5. {(−2, −2), (−1, 1), (1, 1), (2, 2)}, D = {−2, −1, 1, 2}, R = {−2, −1, 1, 2} 6. {(−1, 3), (−1, −1), (1, 2)}, D = {−1, 1}, R = {−1, 2, 3}, no 7. B

© Glencoe/McGraw-Hill   61   Glencoe Pre-Algebra

NAME _____ DATE _____ PERIOD _____

8-2 Linear Equations in Two Variables

(Pages 375–379)

Solving an equation means finding replacement values for the variable that make a true sentence. An equation such as $y = 2x + 3$ is a **linear equation** because its graph is a straight line. The solutions of an equation with two variables are ordered pairs. An equation with two variables usually has an infinite number of solutions.

Graphing Linear Equations	To graph a linear equation with two variables, use the following procedure: • Choose any convenient values for *x*. • Substitute each *x*-value in the equation and solve to find each corresponding *y*-value. Write these solutions as (*x*, *y*) pairs. • Graph at least 3 of the ordered pairs and draw the straight line that passes through them.

Example

Find four solutions for the equation $2x + y = 3$. Then graph the equation.

Choose values for x: −1, 0, 1, 2. Find the corresponding values for y by substituting each x-value in the equation and solving for y.

$2(-1) + y = 3$ $\quad$ $2(0) + y = 3$ $\quad$ $2(1) + y = 3$ $\quad$ $2(2) + y = 3$
$y = 5$ $\qquad\qquad$ $y = 3$ $\qquad\qquad$ $y = 1$ $\qquad\qquad$ $y = -1$

Write these solutions as ordered pairs: (−1, 5), (0, 3), (1, 1), (2, −1).

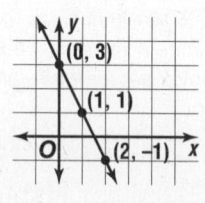

Try This Together

1. Which of these ordered pairs are solutions of $x + y = 8$?
   a. (7, 1) **b.** (−3, 11) **c.** (2, −9) **d.** (4, 4)
   HINT: There may be more than one pair that makes the equation true.

Practice

Which of these ordered pairs is a solution of the given equation?

2. $2x + y = -6$ **a.** (−8, 4) **b.** (−1, −4) **c.** (5, −16) **d.** (9, 1)
3. $-3x = 2y$ **a.** (1, −1) **b.** (7, 10) **c.** (−2, 3) **d.** (5, 5)

Find four solutions for each equation and write them as ordered pairs. Then graph the equation.

4. $y = -3x$ $\qquad$ 5. $y = 2x - 3$ $\qquad$ 6. $y - x = 2$

7. **Standardized Test Practice** Which ordered pair is a solution of the equation $y - x = 7$?
   A (1, 6) $\qquad$ **B** (−1, −6) $\qquad$ **C** (−1, 6) $\qquad$ **D** (1, −6)

Answers: 1. a, b, d 2. b, c 3. c 4–6. See Answer Key. 7. C

© Glencoe/McGraw-Hill $\qquad\qquad$ Glencoe Pre-Algebra

8-3 Graphing Linear Equations Using Intercepts *(Pages 381–385)*

The *x*-intercept for a linear graph is the *x*-coordinate of the point where the graph crosses the *x*-axis and can be found by letting $y = 0$. The *y*-intercept is the *y*-coordinate of the point where the graph crosses the *y*-axis and can be found by letting $x = 0$.

Examples

a. Find the *x*-intercept and the *y*-intercept for the graph of $y = 4x - 2$. Then graph the line.

x-intercept
Let $y = 0$.
$0 = 4x - 2$
$2 = 4x$
$x = \frac{2}{4}$ or $\frac{1}{2}$
x-intercept: $\frac{1}{2}$

y-intercept
Let $x = 0$.
$y = 4(0) - 2$
$y = 0 - 2$ or -2
y-intercept: -2

Graph the ordered pair for each intercept: $\left(\frac{1}{2}, 0\right)$ and $(0, -2)$. Then draw the line that contains them.

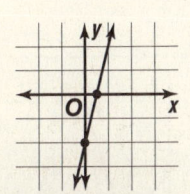

b. Graph the equation $y = -2x + 3$.

x-intercept
Let $y = 0$.
$0 = -2x + 3$
$-3 = -2x$
$\frac{3}{2} = x$
x-intercept: $\frac{3}{2}$ or $1\frac{1}{2}$

y-intercept
Let $x = 0$.
$y = -2(0) + 3$
$y = 0 + 3$ or 3
y-intercept: 3

Graph the ordered pair for each intercept: $\left(1\frac{1}{2}, 0\right)$ and $(0, 3)$. Then draw the line that contains them.

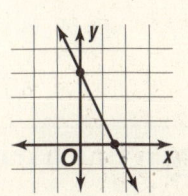

Practice

Find the *x*-intercept and the *y*-intercept for the graph of each equation. Then graph the line.

1. $y = 2x - 3$
2. $y = -x + 1$
3. $y = \frac{2}{3}x - 4$
4. $y = -\frac{1}{2}x + 2$
5. $y = 3x - 2$
6. $y = -2x + 4$

Graph each equation using the slope and *y*-intercept.

7. $y = -x + 3$
8. $y = \frac{1}{3}x + 2$
9. $y = 2x - 1$

10. **Standardized Test Practice** Which of the following is the *x*-intercept for the graph of $y = 3x - 6$?

A -6 B 2 C -2 D 6

Answers: 1–9. See Answer Key. 10. B

© Glencoe/McGraw-Hill   Glencoe Pre-Algebra

NAME _____ DATE _____ PERIOD _____

8-4 Slope (Pages 387–391)

The steepness, or **slope**, of a line can be expressed as the ratio of the vertical change to the horizontal change. The vertical change (or the change up or down) is called the *rise*. The horizontal change (or change right or left) is called the *run*.

Finding the Slope of a Line	You can find the slope of a line by using the coordinates of any two points on the line. • To find the rise, subtract the y-coordinate of the first point from the y-coordinate of the second point. • To find the run, subtract the x-coordinate of the first point from the x-coordinate of the second point. • Write this ratio to find the slope of the line: slope = $\frac{rise}{run}$.

Example

Find the slope of the line that contains the points (25, 2) and (7, 4).

$\frac{rise}{run} = \frac{2nd\ y\text{-coordinate} - 1st\ y\text{-coordinate}}{2nd\ x\text{-coordinate} - 1st\ x\text{-coordinate}}$ Note that order is important.

$= \frac{4 - 2}{7 - (-5)}$

$= \frac{2}{12}$ or $\frac{1}{6}$

Practice

Determine the slope of each line named below.

1. a
2. b
3. c
4. d
5. e
6. f

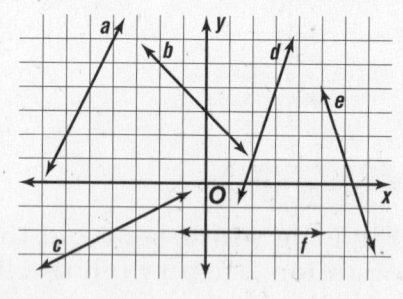

Find the slope of the line that contains each pair of points.

7. $K(3, 9), L(2, 4)$
8. $A(1, 0), B(-3, 1)$
9. $M(8, -6), N(8, 4)$
10. $S(1, -5), T(-3, -4)$
11. $W(1, 6), Z(2, 6)$
12. $P(-4, -5), Q(-3, 7)$

13. **Carpentry** A ladder leans against a building. What is the slope of the ladder if the top of the ladder is 15 feet above the ground and the base of the ladder is 3 feet from the building?

14. **Standardized Test Practice** Find the slope of the line that contains the points $(-3, 2)$ and $(-6, 0)$.

A $\frac{2}{3}$ B $\frac{2}{9}$ C $-\frac{2}{9}$ D $-\frac{2}{3}$

Answers: 1. 2 2. -1 3. $\frac{1}{2}$ 4. 3 5. -3 6. 0 7. 5 8. $-\frac{1}{4}$ 9. no slope 10. $-\frac{1}{4}$ 11. 0 12. 12 13. 5 14. A

© Glencoe/McGraw-Hill   Glencoe Pre-Algebra

NAME _____ DATE _____ PERIOD _____

8-5 Rate of Change (Pages 393–397)

A change in one quantity with respect to another quantity is called a **rate of change.** Any rate of change can be described in terms of slope, or $\frac{\text{change in } y}{\text{change in } x}$. A special type of equation that describes a rate of change is a linear equation in the form of $y = kx$, where $k \neq 0$, and is called **direct variation.** In direct variation we say that y *varies directly with x* or *y varies directly as x*. In the direct variation equation, $y = kx$, k, is the **constant of variation.** The constant of variation in a direct variation equation has the same value as the slope of the graph. For example, $y = 3x$ is a direct variation because it is in the form of $y = kx$. The constant of variation of $y = 3x$ is 3. The slope of the linear graph of $y = 3x$ is 3. All direct variation graphs pass through the origin.

Examples

a. For the equation $y = 2x$, which passes through points (2, 4) and (5, 10), show that the slope and the constant of the variation are equal.

2 is the constant of the variation;
$m = \frac{y_2 - y_1}{x_2 - x_1} = \frac{10 - 4}{5 - 2} = \frac{6}{3} = \frac{2}{1} = 2$

b. Write and solve an equation if y varies directly with x and $y = 40$ when $x = 5$.

$y = kx$ Direct variation form
$40 = k \cdot 5$ Substitute values.
$8 = k$ Divide each side by 5.
Therefore, $y = 8x$.

Practice

Name the constant of variation for each equation. Then determine the slope of the line that passes through the given pair of points.

1. $y = \frac{1}{3}x$; (6, 2), (−9, −3)   2. $y = \frac{-5}{2}x$; (−10, 25), (−2, 5)   3. $y = 13x$; (2, 26), (9, 117)

Write a direct variation equation that relates x and y. Assume that y varies directly with x. Then solve.

4. If $y = -32$ when $x = 4$, find x when $y = 24$.

5. If $y = 15$ when $x = 6$, find x when $y = -25$.

6. **Standardized Test Practice** Which equation is *not* an example of a direct variation?

   A $y = \frac{-7}{3}x + 1$ **B** $y = \frac{5}{16}x$ **C** $y = 14x$ **D** $y = -9x$

Answers: 1. $k = \frac{1}{3}$, $m = \frac{1}{3}$ 2. $k = \frac{-5}{2}$, $m = \frac{-5}{2}$ 3. $k = 13$, $m = 13$ 4. $y = -8x$, $x = -3$ 5. $y = 2.5x$, $x = -10$ 6. A

© Glencoe/McGraw-Hill   65   Glencoe Pre-Algebra

NAME _____ DATE _____ PERIOD ___

8-6 Slope-Intercept Form (Pages 398–401)

Slope-Intercept Form	When a linear equation is written in the form $y = mx + b$, it is in **slope-intercept form**. $y = mx + b$ with arrows pointing to m (slope) and b (y-intercept)

Example

State the slope and the y-intercept of the graph of $y = \frac{2}{3}x - 3$.

$y = \frac{2}{3}x - 3$ Write the original equation.

$y = \frac{2}{3}x + (-3)$ Write the equation in the form $y = mx + b$.

$y = mx + b$ $m = \frac{2}{3}, b = -3$

The slope of the graph is $\frac{2}{3}$, and the y-intercept is -3.

Example

Write the equation $2x + 3y = 5$ in slope-intercept form.

Slope-Intercept Form: $2x + 3y = 5$
$3y = -2x + 5$ Subtract 2x from each side.
$y = -\frac{2}{3}x + \frac{5}{3}$ Divide each side by 3.

Note that in this form we can see that the slope m of the line is $-\frac{2}{3}$, and the y-intercept b is $\frac{5}{3}$.

Practice

State the slope and the y-intercept for the graph of each equation.

1. $y = 2x + 4$ 2. $4x + y = 0$ 3. $5x + 2y = 7$

Write an equation in slope-intercept form of a line with the given slope and y-intercept.

4. $m = 5, b = 5$ 5. $m = 2, b = -7$ 6. $m = -3, b = 0$

Find the slope and y-intercept of the graph of each equation.

7. $7y = x - 10$ 8. $8x - \frac{1}{2}y = -2$ 9. $4(x - 5y) = 9(x + 1)$

10. **Standardized Test Practice** What is the slope-intercept form of an equation for the line that passes through (0, 1) and (3, 37)?
    A $y = 12x - 1$ **B** $y = 12x + 1$ **C** $y = -12x - 1$ **D** $y = -12x + 1$

Answers: 1. 2, 4 2. -4, 0 3. $-\frac{5}{2}, \frac{7}{2}$ 4. $y = 5x + 5$ 5. $y = 2x - 7$ 6. $y = 3x$ 7. $m = \frac{1}{7}, b = -\frac{10}{7}$ 8. $m = 16, b = 4$ 9. $m = -\frac{1}{4}, b = -\frac{9}{20}$ 10. B

© Glencoe/McGraw-Hill   66   Glencoe Pre-Algebra

8-7 Writing Linear Equations *(Pages 404–408)*

To write an equation given the slope and one point	Use $y = mx + b$ for the equation. Replace m with the given slope and the coordinates of the given point for x and y. Solve the equation for the y-intercept, b. Rewrite the equation with the slope for m and the y-intercept for b.
To write an equation given two points	Use the slope formula to calculate m. Choose any of the two given points to use in place of x and y in $y = mx + b$. Replace m with the slope you just calculated. Solve for b. Rewrite the equation with the slope for m and the y-intercept for b.

Examples
Write an equation in slope-intercept form from the given information.

a. The slope is 3 and the line passes through the point (5, 16).

$y = mx + b$	Use slope-intercept form.
$y = 3x + b$	Replace m with the slope.
$16 = 3 \cdot 5 + b$	Replace x and y.
$1 = b$	Solve for b.
$y = 3x + 1$	Rewrite the equation.

b. The line passes through the points (10, −4) and (−7, 13).

$m = \frac{y_2 - y_1}{x_2 - x_1}$	Use the slope formula.
$m = \frac{13 - (-4)}{-7 - 10}$	Substitute.
$m = -1$	Solve for m.
$y = mx + b$	
$-4 = (-1)10 + b$	Substitute m, x, and y.
$6 = b$	Solve for b.
$y = -x + 6$	Rewrite the equation.

Practice
Write an equation in slope-intercept form from the given information.

1. $m = 2$, (6, 1)
2. $m = \frac{1}{2}$, (5, 6.5)
3. $m = 1$, (−5, −7)
4. $m = -\frac{5}{4}$, (−1, 8)

5. (3, 8), (5, 9)
6. (3, −4), (−6, −1)
7. (0, 7), (−2, 3)
8. (−10, 47), (5, −13)

9. **Standardized Test Practice** Which is the correct slope-intercept equation for a line that passes through the points (−15, −47) and (−19, −59)?
   A $y = -3x + 2$
   B $y = 3x + 2$
   C $y = -3x - 2$
   D $y = 3x - 2$

Answers: 1. $y = 2x - 11$ 2. $y = \frac{1}{2}x + 4$ 3. $y = x - 2$ 4. $y = -\frac{5}{4}x + \frac{27}{4}$ 5. $y = \frac{1}{2}x + \frac{13}{2}$ 6. $y = -\frac{1}{3}x - 3$ 7. $y = 2x + 7$ 8. $y = -4x + 7$ 9. D

NAME _____ DATE _____ PERIOD _____

8-8 Best-Fit Lines (Pages 409–413)

When collecting real-life data, the points rarely form a straight line; however, the points may approximate a linear relationship. In this case, a best-fit line may be used. A **best-fit line** is a line that is drawn close to all of the points in the data. In short, it is the line that best fits the points. Best-fit lines help us to write equations for a set of data and predict what may happen if the data continues on the same trend.

Example

The table shows Tisha's height at various ages. Use the information to make a scatter plot, draw a best-fit line, and write an equation for the data.

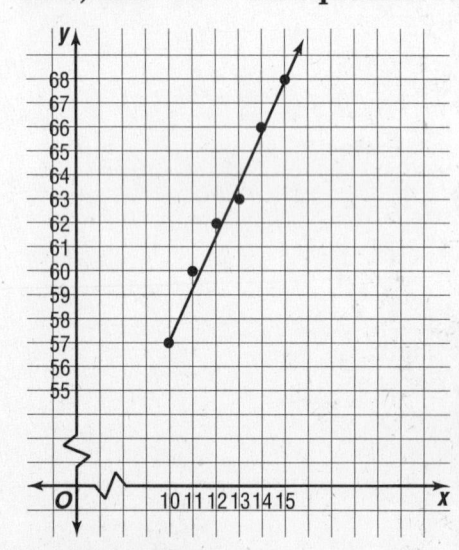

Age	Height in Inches
10	57
11	60
12	62
13	63
14	66.5
15	68

$m = \dfrac{y_2 - y_1}{x_2 - x_1}$ Select two points to find the slope.

$m = \dfrac{62 - 68}{12 - 15}$ $x_1 = 15, y_1 = 68, x_2 = 12, y_2 = 62$

$m = 2$ The slope is 2.

$y = mx + b$ Use the slope-intercept form.

$62 = 2 \cdot 12 + b$ Replace m with 2 and use any point.

$38 = b$ Solve for b.

$y = 2x + 38$ Replace m and b in the equation.

Practice

Use the table that shows the number of goals Pierre scored playing hockey to answer problems 1–4.

Year	Goals
1997	26
1998	24
1999	20
2000	19
2001	15

1. Using the data from 2001 and 1997, find the slope of the line.

2. With your answer from problem 1 and the point (2000, 19), write an equation for the line in slope-intercept form.

3. Using your answer from problem 2, how many goals should Pierre score in 2004?

4. **Standardized Test Practice** What would have been the equation for problem two if the given information was the answer to problem 1 and the point (1998, 24)?

A $y = \dfrac{-11}{4}x + 5518\dfrac{1}{2}$ B $y = \dfrac{11}{4} + 5518\dfrac{1}{2}$ C $y = \dfrac{-11}{4}x - 5518\dfrac{1}{2}$ D $y = \dfrac{11}{4}x - 5518\dfrac{1}{2}$

Answers: 1. $m = \dfrac{-11}{4}$ 2. $y = \dfrac{-11}{4}x + 5519$ 3. 8 4. A

© Glencoe/McGraw-Hill   Glencoe Pre-Algebra

8-9 Solving Systems of Equations (Pages 414–418)

Two equations with the same two variables form a **system of equations**. A **solution** of a system is an ordered pair that is a solution of both equations.

Solutions to Systems of Equations	• When the graphs of two linear equations intersect in exactly one point, the system has exactly one ordered pair as its solution. • When the graphs of two linear equations are parallel, the system has no solution. • When the graphs are the same line, the system has infinitely many solutions.

Example

Use a graph to solve the system of equations $y = x + 1$ and $y = 2x + 3$.

The graph of $y = x + 1$ has an x-intercept of -1 and a y-intercept of 1.
Therefore, two points on this line are $(-1, 0)$ and $(0, 1)$.
The graph of $y = 2x + 3$ has a y-intercept of 3. Thus, one point on this line is $(0, 3)$.
Using the slope of 2 or $\frac{2}{1}$, we find another point on the line at $(1, 5)$.
The graphs of the lines containing each set of points intersect at $(-2, -1)$.
Therefore the solution to the system is $(-2, -1)$.

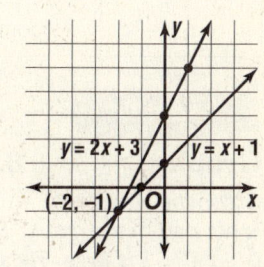

Try These Together

1. Use the graph in PRACTICE to find the solution of the system of equations represented by line a and line b.

2. Use a graph to solve the system of equations $y = -2x + 1$ and $y = x - 2$.

Practice

The graphs of several equations are shown to the right. State the solution of each system of equations.

3. a and c
4. b and d
5. c and the y-axis
6. d and the x-axis
7. a and d
8. b and c

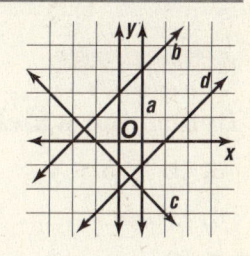

Solve each system of equations by substitution.

9. $y = 2x - 3$
   $y = -x$

10. $y = x - 2$
    $y = 5$

11. $y = -3x + 2$
    $y = -2$

12. **Standardized Test Practice** Which ordered pair is the solution to the system of equations $y = -3x$ and $y = -2x - 4$?
    A $(1, 3)$ B $(2, -6)$ C $(4, -12)$ D $(1, -6)$

Answers: 1. $(1, 3)$ 2. See Answer Key. 3. $(1, -2)$ 4. no solution 5. $(0, -1)$ 6. $(2, 0)$ 7. $(1, -1)$ 8. $(-1.5, 0.5)$ 9. $(1, -1)$ 10. $(3, 5)$ 11. $(8, -2)$ 12. C

NAME _____ DATE _____ PERIOD _____

8-10 Graphing Inequalities *(Pages 419–422)*

The graph of an inequality consists of a dashed or solid **boundary line** and a shaded region. The boundary line is the graph of the equation that corresponds to the inequality. The boundary is dashed if the inequality symbol is $<$ or $>$ to show that these points are not included in the graph. It is solid for $\leq$ or $\geq$ to show that the boundary points are included in the graph.

Graphing Inequalities	• To graph an inequality, first draw the graph of the related equality. This boundary line separates the plane into two regions. If the inequality symbol is $\leq$ or $\geq$, make the boundary line solid; otherwise, it is dashed. • To determine which region to shade as the solution, test a point in each region to see if its coordinates make the inequality true.

Example

Graph $y > -x + 3$.

Graph the equation $y = -x + 3$. Draw a dashed line since the boundary is not part of the graph. The origin (0, 0) is not part of the graph, since $0 > -0 + 3$ is false. Thus, the graph is all points in the region above the boundary. Shade this region.

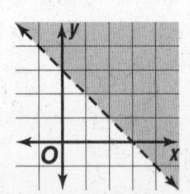

Try These Together

1. Which of the ordered pairs is a solution of $x + y \geq 7$?
   a. $(2, 8)$ **b.** $(-15, 6)$ **c.** $(0, 7)$
   HINT: Replace x and y with the given values to see if they make the inequality true.
2. Graph the inequality $y \geq -2x + 4$.

Practice

Determine which of these ordered pairs is a solution of the inequality.

3. $3x - 5 \leq y$ **a.** $(-2, 4)$ **b.** $(1, -1)$ **c.** $(2, 6)$
4. $y \leq x - 7$ **a.** $(0, -10)$ **b.** $(12, 2)$ **c.** $(-12, -11)$
5. $3x > y - 4$ **a.** $(7, 7)$ **b.** $(-2, 8)$ **c.** $(-1, 0)$

Graph each inequality.

6. $y < x - 7$
7. $y \leq 3x - 5$
8. $y > 1$
9. $y + 4 < x$
10. $x \geq -4$
11. $3x + y \leq 5$

12. **Standardized Test Practice** Which ordered pair is a solution to $2x + y < 5$?
    A $(1, 3)$ **B** $(3, 1)$ **C** $(2, 0)$ **D** $(3, 0)$

Answers: 1. a, c   2. See Answer Key.   3. a, b   4. a, b   5. a, b   6–11. See Answer Key.   12. C

© Glencoe/McGraw-Hill — 70 — Glencoe Pre-Algebra

Chapter Review
Breakfast Riddle

What is on the breakfast menu for the school cafeteria this morning? To find out what they are serving, work the following problems. Look for your answer in the right column of the box. Use the corresponding letters in the left column to fill in the blanks below.

___ ___ ___ ___ ___ ___ ___ ___ ___ ___ ___

(Fill in the blanks in the order of the questions.)

Start by graphing the following three equations on the same coordinate plane: $f(x) = x + 3$, $g(x) = -\frac{1}{2}x$, and $x = 4$.

1. Which of the three graphs represent functions?

2. What is the x-intercept of the graph of $f(x)$?

3. What is the y-intercept of the graph of $g(x)$?

4. What is the slope of the graph of $f(x)$?

5. For which of the three equations is (1, 4) a solution?

6. Find $g(8)$.

D	−1
N	4
S	−4
LE	$y = 0$
NI	$y = 3$
MA	$x = 0$
NO	$x = 3$
OI	$x = -3$
PR	(4, −2)
DE	1
ST	(2, 1)
SO	All 3 graphs
B	$f(x)$ and $g(x)$
NO	$f(x)$ and $x = 4$
GG	$f(x)$ only
TO	$g(x)$ only

Answers are located in the Answer Key.

NAME _____ DATE _____ PERIOD _____

9-1 Squares and Square Roots (Pages 436–440)

A **square root** is one of two equal factors of a number. For example, the square root of 25 is 5 because $5 \cdot 5$ or 5^2 is 25. Since $-5 \cdot (-5)$ is also 25, -5 is also a square root of 25.

Definition of Square Root	The square root of a number is one of its two equal factors. If $x^2 = y$, then x is a square root of y. The symbol $\sqrt{}$ is called the **radical sign** and is used to indicate a nonnegative square root. $\sqrt{25}$ indicates the nonnegative square root of 25, so $\sqrt{25} = 5$. $-\sqrt{25}$ indicates the negative square root of 25, so $-\sqrt{25} = -5$.

Numbers like 25, 49, and 64 are called **perfect squares** because their nonnegative square roots are whole numbers. Numbers that are not perfect squares do not have whole number square roots. You can use perfect squares to estimate the square root of a number that is not a perfect square. See Example B below.

Examples

a. Find $\sqrt{64}$.
The symbol $\sqrt{64}$ represents the nonnegative square root of 64. Since $8 \cdot 8 = 64$, $\sqrt{64} = 8$.

b. Find the best integer estimate for $\sqrt{44}$.
Locate the closest perfect squares to 44. They are 36 and 49. Because $36 < 44 < 49$, you know that $\sqrt{36} < \sqrt{44} < \sqrt{49}$, or $6 < \sqrt{44} < 7$. So, $\sqrt{44}$ is between 6 and 7. Since 44 is closer to 49 than to 36, then $\sqrt{44}$ is closer to 7 than 6. The best integer estimate for $\sqrt{44}$ is 7.

Practice

Find each square root.

1. $\sqrt{16}$
2. $-\sqrt{36}$
3. $\sqrt{36}$
4. $\sqrt{121}$
5. $\sqrt{225}$
6. $-\sqrt{900}$

Find the best integer estimate for each square root. Then check your estimate with a calculator.

7. $\sqrt{45}$
8. $\sqrt{29}$
9. $\sqrt{5}$
10. $\sqrt{640}$
11. $-\sqrt{250}$
12. $-\sqrt{57}$
13. $\sqrt{10}$
14. $\sqrt{6.2}$
15. $\sqrt{2}$

16. **Art Framing** A man has a favorite square picture he wants to frame using a mat technique. He knows the area of the picture is 144 in^2.
    a. How would he find the length of the sides of the picture for the mat?
    b. What is the length of each side?

17. **Standardized Test Practice** Find $\sqrt{529}$.
    A 23 B 25 C 52 D 529

Answers: 1. 4 2. −6 3. 6 4. 11 5. 15 6. −30 7. 7 8. 5 9. 2 10. 25 11. −16 12. −8 13. 3 14. 2 15. 1 16a. Find the square root of 144 in^2. 16b. 12 in. 17. A

NAME _____ DATE _____ PERIOD _____

9-2 The Real Number System (Pages 441–445)

You know that rational numbers can be expressed as $\frac{a}{b}$, where a and b are integers and $b \neq 0$. Rational numbers may also be written as decimals that either terminate or repeat. However, there are many numbers (for example, square roots of whole numbers that are not perfect squares) that neither terminate nor repeat. These are called **irrational numbers**.

Definition of an Irrational Number	An irrational number is a number that cannot be expressed as $\frac{a}{b}$, where a and b are integers and b does not equal 0.

The set of rational numbers and the set of irrational numbers make up the set of real numbers. The Venn diagram at the right shows the relationships among the number sets.

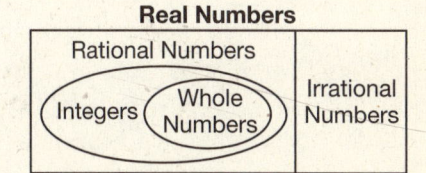

Examples

a. Determine whether 0.121231234 … is rational or irrational.
This decimal does not repeat nor terminate. It does have a pattern to it, but there is no exact repetition. This is an irrational number.

b. Solve $h^2 = 50$. Round your answer to the nearest tenth.
$h^2 = 50$
$h = \sqrt{50}$ or $h = -\sqrt{50}$ Take the square root of each side.
$h \approx 7.1$ or $h \approx -7.1$ Use a calculator.

Practice

Name the sets of numbers to which each number belongs: the whole numbers, the integers, the rational numbers, the irrational numbers, and/or the real numbers.

1. $\frac{3}{4}$
2. 12
3. 0.008
4. $\sqrt{13}$
5. 16.7
6. $-\sqrt{7}$

Solve each equation. Round decimal answers to the nearest tenth.

7. $a^2 = 81$
8. $n^2 = 54$
9. $37 = m^2$
10. $p^2 = 6$
11. $18 = w^2$
12. $x^2 = 99$
13. $k^2 = 5$
14. $s^2 = 82$
15. $61 = b^2$

16. **Physics** If you drop an object from a tall building, the distance d in feet that it falls in t seconds can be found by using the formula $d = 16t^2$. How many seconds would it take a dropped object to fall 64 feet?

17. **Standardized Test Practice** Find the positive solution of $y^2 = 254$. Round to the nearest tenth.
    A 15.4   B 15.6   C 15.7   D 15.9

Answers: 1. rational, real 2. whole number, integer, rational, real 3. rational, real 4. irrational, real 5. rational, real 6. irrational, real 7. 9, −9 8. 7.3, −7.3 9. 6.1, −6.1 10. 2.4, −2.4 11. 4.2, −4.2 12. 9.9, −9.9 13. 2.2, −2.2 14. 9.1, −9.1 15. 7.8, −7.8 16. 2s 17. D

NAME _____ DATE _____ PERIOD _____

9-3 Angles (Pages 447–451)

Common Geometric Figures and Terms

point A	vertex M 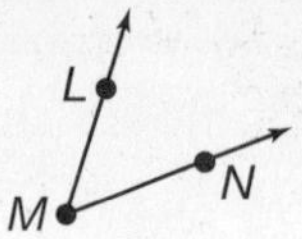	An **acute** angle measures between 0° and 90°.
ray FG or $\vec{FG}$ 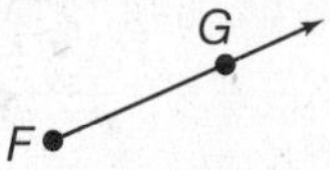	The **sides** of ∠LMN are $\vec{ML}$ and $\vec{MN}$. 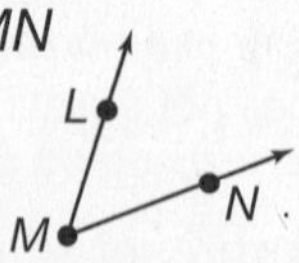	A **right** angle measures 90°.
line BC or $\overleftrightarrow{BC}$ or line ℓ 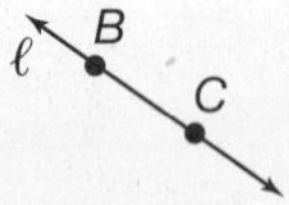	Angles are measured in **degrees** (°) using a protractor.	An **obtuse** angle measures between 90° and 180°.
angle LMN or ∠LMN 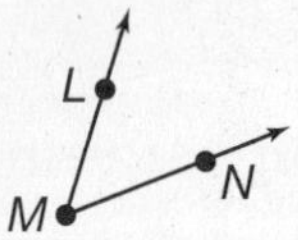	A **protractor** is used to measure angles. 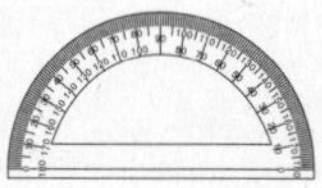	A **straight** angle measures 180°.

Example

Use a protractor to measure ∠TQR.

Place the protractor so the center is at the vertex Q and the straightedge aligns with side $\vec{QR}$. Use the scale that begins with 0 (on $\vec{QR}$). Read where side $\vec{QT}$ crosses this scale. The measure of ∠TQR is 120 degrees. In symbols, this is written m∠TQR = 120°.

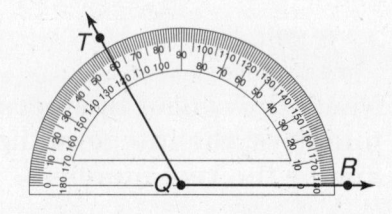

Practice

Use a protractor to find the measure of each angle. Then classify each angle as *acute, obtuse, right,* or *straight*.

1. ∠DAB
2. ∠HAE
3. ∠HAD
4. ∠BAC
5. ∠CAF
6. ∠GAE
7. ∠HAB
8. ∠GAC

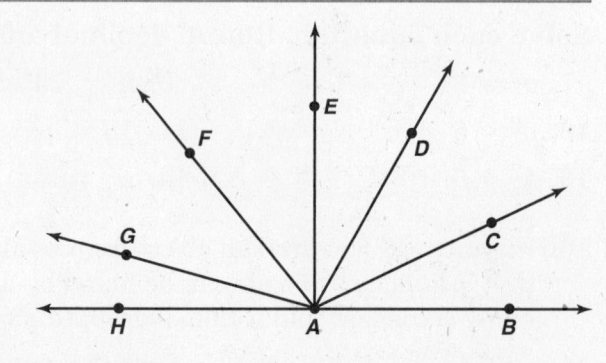

9. **Standardized Test Practice** What is the vertex of ∠KLM?

   A point K   B point L   C point M   D point KLM

Answers: 1. 55°; acute 2. 90°; right 3. 120°; obtuse 4. 25°; acute 5. 105°; obtuse 6. 75°; acute 7. 180°; straight 8. 140°; obtuse

© Glencoe/McGraw-Hill 74 Glencoe Pre-Algebra

NAME _____ DATE _____ PERIOD _____

9-4 Triangles (Pages 453–457)

A triangle is formed by three line segments that intersect only at their endpoints. The sum of the measures of the angles of a triangle is 180°.

Classifying Triangles	You can classify a triangle by its *angles*. • An acute triangle has three acute angles. • A right triangle has one right angle. • An obtuse triangle has one obtuse angle. You can classify a triangle by the *number of congruent sides*. • An equilateral triangle has 3 congruent sides. • An isosceles triangle has at least two congruent sides. • A scalene triangle has no congruent sides.

Try These Together
Use the figure in PRACTICE below to answer these questions.

1. Find $m\angle 1$ if $m\angle 2 = 50°$ and $m\angle 3 = 55°$.
2. Find $m\angle 1$ if $m\angle 2 = 65°$ and $m\angle 3 = 55°$.

HINT: The sum of all angle measures in a triangle is 180°.

Practice

Use the figure at the right to solve each of the following.

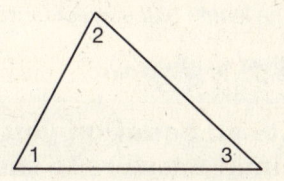

3. Find $m\angle 1$ if $m\angle 2 = 52°$ and $m\angle 3 = 69°$.
4. Find $m\angle 1$ if $m\angle 2 = 62°$ and $m\angle 3 = 44°$.
5. Find $m\angle 1$ if $m\angle 2 = 71°$ and $m\angle 3 = 22°$.
6. Find $m\angle 1$ if $m\angle 2 = 90°$ and $m\angle 3 = 30°$.

First classify each triangle as acute, right, or obtuse. Then classify each triangle as scalene, isosceles, or equilateral.

7. 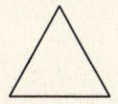 8. 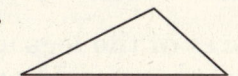 9. 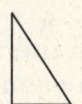 10.

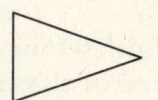

11. **Food** Samantha likes her grilled cheese sandwiches cut in half diagonally. Classify the triangles that come from slicing a square diagonally. Are they acute, right, or obtuse? Are they scalene, isosceles, or equilateral?

12. **Standardized Test Practice** Find the measure of $\angle A$. 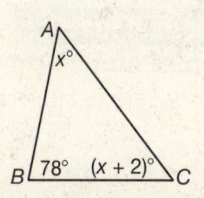
    A 60°   B 52°
    C 50°   D 48°

Answers: 1. 75° 2. 60° 3. 59° 4. 74° 5. 87° 6. 60° 7. acute; equilateral 8. obtuse; scalene 9. right; scalene 10. acute; isosceles 11. right; isosceles 12. C

NAME _____ DATE _____ PERIOD _____

9-5 The Pythagorean Theorem (Pages 460–464)

The **Pythagorean Theorem** describes the relationship between the **legs** of a right triangle, the sides that are adjacent to the right angle, and the **hypotenuse**, the side opposite the right angle.

Pythagorean Theorem	In a right triangle, the square of the length of the hypotenuse is equal to the sum of the squares of the lengths of the legs, or $c^2 = a^2 + b^2$.

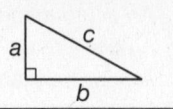

You can use the Pythagorean Theorem to find the length of any side of a right triangle as long as you know the lengths of the other two sides. You can also use the Pythagorean Theorem to determine if a triangle is a right triangle.

Examples

a. If a right triangle has legs with lengths of 9 cm and 12 cm, what is the length of the hypotenuse?

$c^2 = a^2 + b^2$ Pythagorean Theorem
$c^2 = 9^2 + 12^2$ Replace a with 9 and b with 12.
$c^2 = 225$ Simplify.
$c = \sqrt{225}$ Take the square root of each side.
$c = 15$ Simplify.

The length of the hypotenuse is 15 cm.

b. Is a triangle with side lengths of 6 m, 9 m, and 12 m a right triangle? Remember, the hypotenuse is the longest side.

$c^2 = a^2 + b^2$ Pythagorean Theorem
$12^2 \stackrel{?}{=} 6^2 + 9^2$ $a = 6, b = 9, c = 12$
$144 \stackrel{?}{=} 36 + 81$ Multiply.
$144 \neq 117$ Add.

The triangle is not a right triangle.

Practice

Write an equation you could use to solve for *x*. Then solve. Round decimal answers to the nearest tenth.

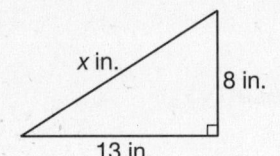

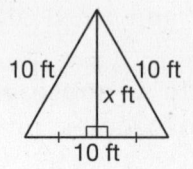

1. [triangle with legs x in. and 8 in., hypotenuse 13 in.]

2. [isosceles triangle with two sides 10 ft, base 10 ft, height x ft]

In a right triangle, if *a* and *b* are the measures of the legs and *c* is the measure of the hypotenuse, find each missing measure. Round decimal answers to the nearest tenth.

3. $a = 5, b = 6$ 4. $c = 14, a = 8$ 5. $a = 9, c = 18$ 6. $a = 7, b = 7$

7. The measurements of three sides of a triangle are 12 feet, 13 feet, and 5 feet. Is this a right triangle? Explain.

8. **Standardized Test Practice** In a right triangle, the legs have lengths 12 centimeters and 15 centimeters. What is the length of the hypotenuse? Round to the nearest tenth.
   A 1.93 cm   B 19.0 cm   C 19.2 cm   D 190 cm

Answers: 1. 15.3 in. 2. 8.7 ft 3. 7.8 4. 11.5 5. 15.6 6. 9.9 7. yes; $5^2 + 12^2 = 13^2$ 8. C

© Glencoe/McGraw-Hill   76   Glencoe Pre-Algebra

NAME _____ DATE _____ PERIOD _____

9-6 The Distance and Midpoint Formulas

(Pages 466–470)

Sometimes it is necessary to study line segments on the coordinate plane. A **line segment**, or a part of a line, contains two endpoints. The coordinates of these endpoints can help us find the length and the **midpoint**, or the point that is halfway between the two endpoints, of the line segment. We can calculate the length of a line segment by using the **Distance Formula**, and we can calculate the midpoint of a line segment by using the **Midpoint Formula**.

The Distance Formula	To calculate the distance d of a line segment with endpoints (x_1, y_1) and (x_2, y_2) use the formula $d = \sqrt{(x_2 - x_1)^2 + (y_2 - y_1)^2}$.
The Midpoint Formula	To calculate the midpoint of a line segment with endpoints (x_1, y_1) and (x_2, y_2) use the formula $\left(\dfrac{x_1 + x_2}{2}, \dfrac{y_1 + y_2}{2}\right)$.

Examples

a. Find the distance between (2, 3) and (6, 8).

Let $x_1 = 2$, $x_2 = 6$, $y_1 = 3$, and $y_2 = 8$.

$d = \sqrt{(x_2 - x_1)^2 + (y_2 - y_1)^2}$
$d = \sqrt{(6 - 2)^2 + (8 - 3)^2}$
$d = \sqrt{4^2 + 5^2}$
$d = \sqrt{16 + 25}$
$d = \sqrt{41}$ or 6.4 units

b. Find the midpoint of (5, 1) and (−1, 5).
Let $x_1 = 5$, $x_2 = -1$, $y_1 -1$, and $y_1 = 5$.

$\left(\dfrac{x_1 + x_2}{2}, \dfrac{y_1 + y_2}{2}\right)$ Midpoint Formula
$\left(\dfrac{5 + -1}{2}, \dfrac{1 + 5}{2}\right)$ Substitute.
$\left(\dfrac{4}{2}, \dfrac{6}{2}\right)$ Add.
(2, 3) is the midpoint

Practice

Find the distance between each pair of points. Round answers to the nearest hundredth.

1. (4, 6), (1, 5)   **2.** (15, 4), (10, 10)   **3.** (−7, −2), (11, 3)

Find the midpoint of the given points.

4. (7, −5), (9, −1)   **5.** (−8, 4), (3, −4)   **6.** (−1.8, 1.9), (1.1, 2.8)

7. **Standardized Test Practice** What is the midpoint of the line segment with endpoints (5, −1) and (−9, 7)?
A (2, −3)   B (−2, 3)   C (3, −2)   D (−3, 2)

Answers: 1. 3.16 units 2. 7.81 units 3. 18.68 units 4. (8, −3) 5. (−2.5, 0) 6. (−0.35, 2.35) 7. B

© Glencoe/McGraw-Hill   Glencoe Pre-Algebra

9-7 Similar Triangles and Indirect Measurement (Pages 471–475)

Figures that have the same shape but not necessarily the same size are **similar figures**. The symbol ~ means *is similar to*.

Similar Triangles	• If two triangles are similar, then the corresponding angles are congruent. If the corresponding angles of two triangles are congruent, then the triangles are similar.
	• If two triangles are similar, then their corresponding sides are proportional. If the corresponding sides of two triangles are proportional, then the triangles are similar.

Example

If $\triangle MNP \sim \triangle KLQ$, find the value of x.
Write a proportion using the known measures.

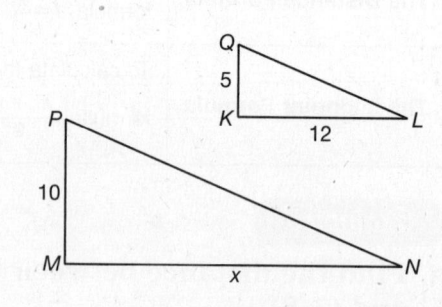

$\frac{QK}{KL} = \frac{PM}{MN}$ Corresponding sides are proportional.

$\frac{5}{12} = \frac{10}{x}$ Substitute.

$5x = 12 \cdot 10$ Find the cross products.

$5x = 120$ Multiply.

$x = 24$ The measure of $\overline{MN}$ is 24.

Practice

$\triangle ABC \sim \triangle DEF$. Use the two triangles to solve each of the following.

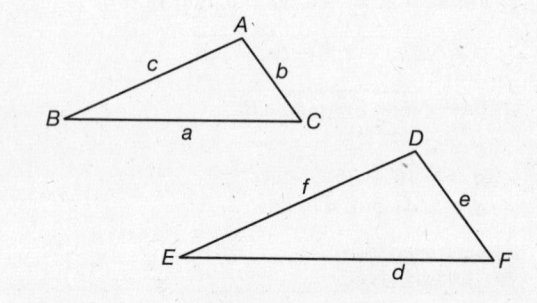

1. Find b if $e = 4$, $a = 9$, and $d = 12$.
2. Find c if $f = 9$, $b = 8$, and $e = 12$.
3. Find d if $a = 6$, $f = 7$, and $c = 5$.
4. Find e if $d = 30$, $a = 10$, and $b = 6$.

5. **Standardized Test Practice** Ancient Greeks used similar triangles to measure the height of a column. They measured the shadows of a column and a smaller object at the same time of day. Then they measured the height of the smaller object and solved for the height of the column. In the picture to the right, use the length of the shadows and the height of the smaller object to solve for the height of the flagpole.

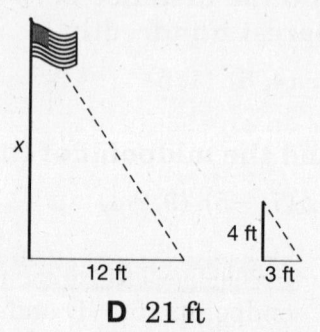

   A 15 ft   B 16 ft   C 20 ft   D 21 ft

Answers: 1. 3 2. 6 3. 8.4 4. 18 5. B

© Glencoe/McGraw-Hill   Glencoe Pre-Algebra

NAME _____ DATE _____ PERIOD _____

9-8 Sine, Cosine, and Tangent Ratios

(Pages 477–481)

Trigonometry is the study of triangle measurement. The ratios of the measures of the sides of a right triangle are called **trigonometric ratios**. Three common trigonometric ratios are defined below.

Definition of Trigonometric Ratios	If △ABC is a right triangle and A is an acute angle, sine of ∠A = $\dfrac{\text{measure of the leg opposite } \angle A}{\text{measure of the hypotenuse}}$ cosine of ∠A = $\dfrac{\text{measure of the leg adjacent to } \angle A}{\text{measure of the hypotenuse}}$ tangent of ∠A = $\dfrac{\text{measure of the leg opposite } \angle A}{\text{measure of the leg adjacent to } \angle A}$ Symbols: $\sin A = \dfrac{a}{c}$, $\cos A = \dfrac{b}{c}$, $\tan A = \dfrac{a}{b}$

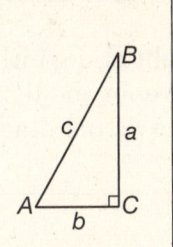

Examples

a. Find sin K to the nearest thousandth.

$\sin K = \dfrac{\text{measure of leg opposite } \angle K}{\text{measure of hypotenuse}}$

$= \dfrac{4}{5}$ or 0.8

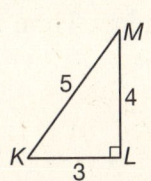

b. Use a calculator to find the measure of ∠A given that sin A = 0.7071.

Enter 0.7071 and the press the sin⁻¹ key (you may have to press INV or 2nd and then the sin key to get sin⁻¹) The calculator should then display 44.999451. Rounded to the nearest degree, the measure of ∠A is 45°.

Practice

For each triangle, find sin A, cos A, and tan A to the nearest thousandth.

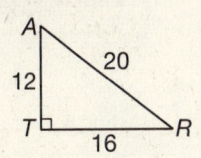

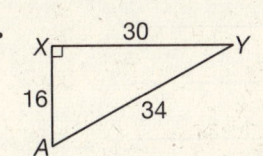

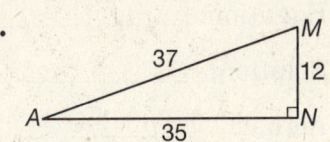

1. (triangle A, T, R with sides 12, 20, 16)
2. (triangle X, Y, A with sides 30, 34, 16)
3. (triangle A, M, N with sides 37, 12, 35)

Use a calculator to find each ratio to the nearest ten thousandth.

4. cos 43°   5. sin 26°   6. sin 36°   7. tan 68°   8. cos 75°   9. tan 29°

Use a calculator to find the angle that corresponds to each ratio. Round answers to the nearest degree.

10. sin X = 0.5   11. sin B = 0.669   12. tan K = 1.881

13. **Sports** A skateboarder builds a ramp to perform jumps. If the ramp is 5 feet long and 3 feet high, what angle does it make with the ground?

14. **Standardized Test Practice** Use a calculator to find sin 56° to the nearest ten thousandth.

   A 0.5600   B 0.8290   C 1.6643   D 88.9770

Answers: 1. 0.8; 0.6; 1.333 2. 0.882; 0.471; 1.875 3. 0.324; 0.946; 0.343 4. 0.7314 5. 0.4384 6. 0.5878 7. 2.4751 8. .2588 9. 0.5543 10. 30° 11. 42° 12. 62° 13. about 31° 14. B

© Glencoe/McGraw-Hill   Glencoe Pre-Algebra

NAME _____ DATE _____ PERIOD _____

9 Chapter Review
Household Hypotenuses

You will need a tape measure, measuring tape, or yardstick and a parent or friend to help. Convert measurements that are fractions into decimals. Round all solutions to the nearest hundredth. (Example: $2\frac{1}{8}$ inches ≈ 2.13 inches)

1. Measure the height and width of your front door. Write an equation and solve for the length of the diagonal of the door. Then measure the actual diagonal and compare.

    Equation: _____

    Solution: _____

    Actual: _____

2. Measure the width and height of a window. Write an equation and solve for the length of the diagonal of the window. Then measure the actual diagonal and compare.

    Equation: _____

    Solution: _____

    Actual: _____

3. Measure the width and diagonal of your TV screen. Write an equation and solve for the height of the TV screen. Then measure the actual height and compare.

    Equation: _____

    Solution: _____

    Actual: _____

4. Measure the length and the diagonal of the top of the mattress on your bed. Write an equation and solve for the width of the mattress. Then measure the actual width and compare.

    Equation: _____

    Solution: _____

    Actual: _____

5. Give some reasons why your solutions are different from the actual measurements.

Answers are located in the Answer Key.

© Glencoe/McGraw-Hill   Glencoe Pre-Algebra

NAME _____ DATE _____ PERIOD _____

10-1 Line and Angle Relationships *(Pages 492–497)*

When two lines intersect, the two pairs of "opposite" angles formed are called **vertical angles**. Vertical angles are always **congruent**, meaning they have the same measure. In Figure 1, ∠1 and ∠3 are vertical angles, so ∠1 is congruent to ∠3.

Two lines in a plane that never intersect are **parallel lines**. In Figure 1, two parallel lines, n and m, are intersected by a third line, p, called a **transversal**. Since n ∥ m, the following statements are true.
∠5 and ∠3 are a pair of congruent **alternate interior angles**.
∠2 and ∠8 are a pair of congruent **alternate exterior angles**.
∠1 and ∠5 are a pair of congruent **corresponding angles**.

Lines that intersect to form a right angle are **perpendicular lines**. In Figure 2, lines a and b are perpendicular. ∠10 and ∠11 are **complementary angles** since the measure of ∠10 plus the measure of ∠11, m∠10 + m∠11, is 90°. ∠9 and ∠13 are **supplementary angles** since m∠9 + m∠13 = 180°

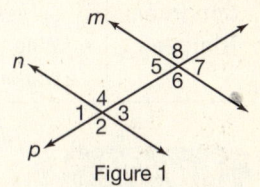

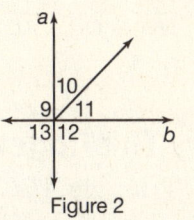

Examples

a. Use Figure 1 to name another pair of vertical angles, congruent alternate interior angles, alternate exterior angles, and corresponding angles.
∠6 and ∠8; ∠4 and ∠6; ∠1 and ∠7; ∠3 and ∠7

b. In Figure 2, if m∠10 = 48°, find m∠11.
∠10 and ∠11 are complementary angles.
m∠10 + m∠11 = 90°
48 + m∠11 = 90°   Substitute.
m∠11 = 42°   Subtract 48 from each side.

Practice

Find the value of x in each figure.

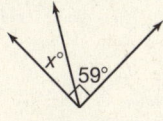

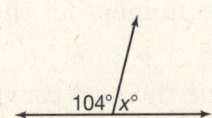

1. (x°, 59°)

2. (104°, x°)

In the figure at the right, line m is parallel to line n. If the measure of ∠1 is 83°, find the measure of each angle.

3. ∠4 **4.** ∠2 **5.** ∠3 **6.** ∠7 **7.** ∠5 **8.** ∠6

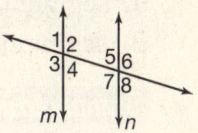

9. Plumbing If a shower head comes out of the wall at an angle of 125°, what is the measure of the other angle between the shower head and the wall?

10. Standardized Test Practice Suppose that ∠F and ∠G are complementary. Find m∠F if m∠G = 11°.
  A 179°   **B** 169°   **C** 79°   **D** 69°

Answers: 1. 31° 2. 76° 3. 83° 4. 97° 5. 97° 6. 97° 7. 83° 8. 97° 9. 55° 10. C

© Glencoe/McGraw-Hill   81   Glencoe Pre-Algebra

NAME _____ DATE _____ PERIOD _____

10-2 Congruent Triangles *(Pages 500–504)*

Figures that have the same size and shape are **congruent**. Parts of congruent triangles that match are called **corresponding parts**.

Congruent Triangles	• If two triangles are congruent, their corresponding sides are congruent and their corresponding angles are congruent. • When you write that triangle ABC is congruent to (≅) triangle XYZ, the corresponding vertices are written in order: △ABC ≅ △XYZ. This means that vertex A corresponds to vertex X, and so on.

Example

△PQR ≅ △JKL. **Write three congruence statements for corresponding sides.**

$\overline{PQ} \cong \overline{JK}$ $\overline{QR} \cong \overline{KL}$ $\overline{RP} \cong \overline{LJ}$

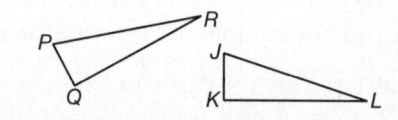

Try This Together

1. Triangle ABC is congruent to triangle DEF.
   a. Name the congruent angles.
   b. Name the congruent sides.

   HINT: Start with the shortest side of each triangle.

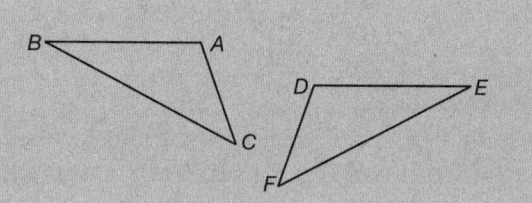

Practice

2. The two triangles at the right are congruent.
   a. Name the congruent angles.
   b. Name the congruent sides.
   c. Write a congruence statement for the triangles themselves.

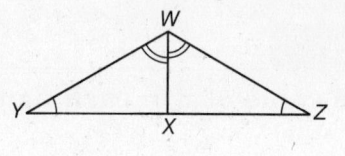

3. If △PQR ≅ △DOG, name the part congruent to each angle or segment given. (*Hint*: Make a drawing.)
   a. segment PQ
   b. segment PR
   c. ∠O
   d. segment OG
   e. ∠G
   f. ∠P

4. **Standardized Test Practice** Which pair of objects best illustrates congruence?
   A a 10 oz can and an 8 oz can
   B two houses that have the same square footage
   C a baseball and softball
   D a CD-Rom and a music CD

Answers: 1. a. ∠A ≅ ∠D; ∠B ≅ ∠E, ∠C ≅ ∠F b. segment AB ≅ segment DE; segment AC ≅ segment DF; segment BC ≅ segment EF 2. a. ∠Y ≅ ∠Z, ∠YWX ≅ ∠ZWX, ∠YXW ≅ ∠ZXW b. segment WY ≅ segment WZ; segment WX ≅ segment WX; segment XY ≅ segment XZ c. △WXY ≅ △WXZ 3. a. segment DO b. segment DG c. ∠O d. segment QR e. ∠R f. ∠D 4. D

© Glencoe/McGraw-Hill   82   Glencoe Pre-Algebra

NAME _____ DATE _____ PERIOD _____

10-3 Transformations on the Coordinate Plane (Pages 506–511)

Transformations are movements of geometric figures, such as translations, rotations, and reflections.

Transformations	• A **translation** is a slide where the figure is moved horizontally or vertically or both. • A **rotation** is a turn around a point. • A **reflection** is a flip of the figure over a line. The transformed figure is the mirror image of the original. The original figure and its reflection form a **symmetric** figure. The line where you placed the mirror is called a line of symmetry. Each line of symmetry separates a figure into two congruent parts.

Example

Which type of transformation does this picture show?
The figure has been rotated around the origin. This is a rotation.

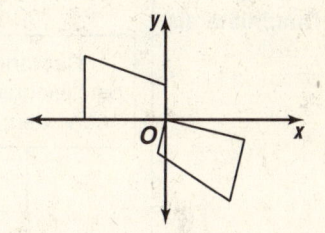

Practice

Tell whether each geometric transformation is a *translation*, a *reflection*, or a *rotation*.

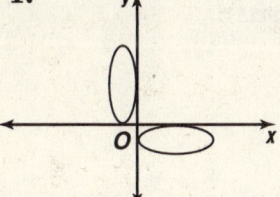

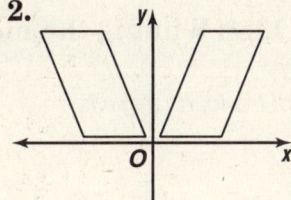

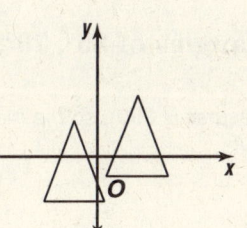

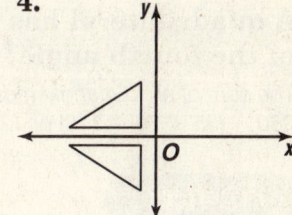

1.
2.
3.
4.

Trace each figure. Draw all lines of symmetry.

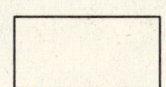

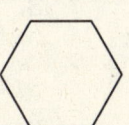

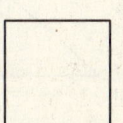

5.
6.
7.
8.

9. **Carpentry** How many ways can you slice a rectangular block of wood into two smaller congruent rectangular blocks? You may want to look at a cereal box to help visualize the situation.

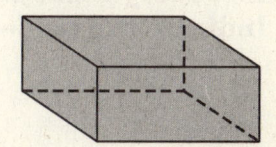

10. **Standardized Test Practice** Which of the following has exactly one line of symmetry?

A   B   C   D

Answers: 1. rotation 2. reflection 3. translation 4. reflection 5–8. See Answer Key. 9. 3 10. D

© Glencoe/McGraw-Hill   83   Glencoe Pre-Algebra

NAME _____ DATE _____ PERIOD _____

10-4 Quadrilaterals (Pages 513–517)

A **quadrilateral** is a closed figure formed by four line segments that intersect only at their endpoints. The sum of the measures of the angles of a quadrilateral is 360°.

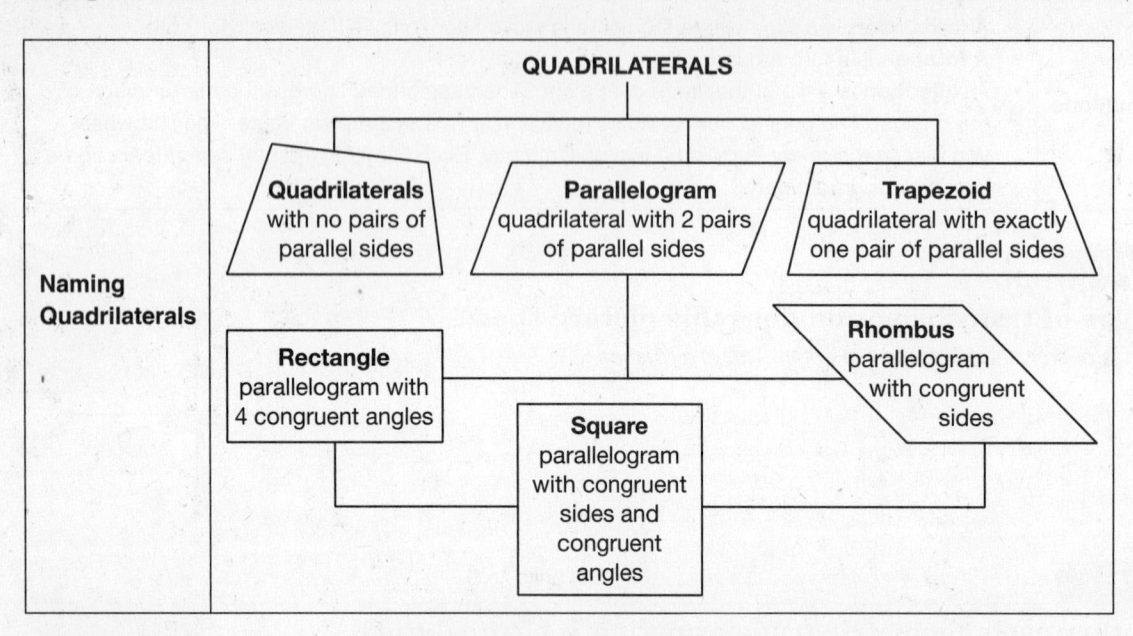

Naming Quadrilaterals

QUADRILATERALS
- **Quadrilaterals** with no pairs of parallel sides
- **Parallelogram** quadrilateral with 2 pairs of parallel sides
- **Trapezoid** quadrilateral with exactly one pair of parallel sides
- **Rectangle** parallelogram with 4 congruent angles
- **Rhombus** parallelogram with congruent sides
- **Square** parallelogram with congruent sides and congruent angles

Example

A quadrilateral has angles of 35°, 79°, and 118°. What is the measure of the fourth angle?

The sum of all four angle measures is 360°, so the measure of the fourth angle is 360 − (35 + 79 + 118) or 128°.

Practice

Find the value of x. Then find the missing angle measures.

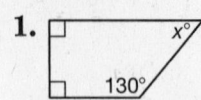

1. (figure with right angle, $x°$, 130°)
2. (parallelogram with 140°, $x°$, $x°$, 140°)
3. (figure with $(x + 5)°$, 135°, $x°$, right angle)

List every name that can be used to describe each quadrilateral. Indicate the name that best describes the quadrilateral.

4. (quadrilateral) 5. (square) 6. (parallelogram) 7. (rhombus) 8. (trapezoid)

9. **Standardized Test Practice** Determine which statement is false.
   A A rhombus is a parallelogram.
   B A rectangle is a parallelogram.
   C A square is a rectangle.
   D A trapezoid is a parallelogram.

Answers: 1. 50; 50° 2. 40; 40° 3. 65; 65°;70° 4. quadrilateral 5. quadrilateral, parallelogram, rhombus, rectangle, square; square 6. quadrilateral, parallelogram 7. quadrilateral, parallelogram, rhombus; rhombus 8. quadrilateral, trapezoid; trapezoid 9. D

© Glencoe/McGraw-Hill   84   Glencoe Pre-Algebra

NAME _____ DATE _____ PERIOD _____

10-5 Area: Parallelograms, Triangles, and Trapezoids (Pages 520–525)

When you find the area of a parallelogram, triangle, or trapezoid, you must know the measure of the **base** and the height. The height is the length of an **altitude**. Use the table below to help you define the bases and heights (altitudes), and find the areas of parallelograms, triangles, and trapezoids.

Parallelogram	Base: any side of the parallelogram Height: the length of an altitude, which is a segment perpendicular to the base, with endpoints on the base and the side opposite the base Area: If a parallelogram has a base of b units and a height of h units, then the area A is $b \cdot h$ square units or $A = b \cdot h$.
Triangle	Base: any side of the triangle Height: the length of an altitude, which is a line segment perpendicular to the base from the opposite vertex Area: If a triangle has a base of b units and a height of h units, then the area A is $\frac{1}{2} b \cdot h$ square units or $A = \frac{1}{2} b \cdot h$.
Trapezoid	Bases: the two parallel sides Height: the length of an altitude, which is a line segment perpendicular to both bases, with endpoints on the base lines Area: If a trapezoid has bases of a units and b units and a height of h units, then the area A of the trapezoid is $\frac{1}{2} \cdot h \cdot (a + b)$ square units or $A = \frac{1}{2} h(a + b)$.

Practice

Find the area of each figure.

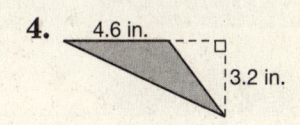

1. parallelogram, 11 ft, 6 ft
2. triangle, 17 cm, 12 cm
3. trapezoid, 9 in., 5 in., 5 in.
4. triangle, 4.6 in., 3.2 in.

Find the area of each figure described below.

5. trapezoid: height, 3 in.; bases, 4 in. and 5 in.
6. triangle: base, 9 cm; height, 8 cm
7. parallelogram: base, 7.25 ft; height, 8 ft
8. triangle: base, 0.3 m; height, 0.6 m

9. **Standardized Test Practice** What is the area of a trapezoid whose bases are 4 yards and 2 yards and whose height is 10 yards?
   A 24 yd²   B 30 yd²   C 60 yd²   D 80 yd²

Answers: 1. 66 ft² 2. 102 cm² 3. 35 in² 4. 7.36 in² 5. 13.5 in² 6. 36 cm² 7. 58 ft² 8. 0.09 m² 9. B

© Glencoe/McGraw-Hill — 85 — Glencoe Pre-Algebra

NAME _____ DATE _____ PERIOD _____

10-6 Polygons (Pages 527–531)

A **polygon** is a simple, closed figure in a plane that is formed by three or more line segments, called **sides**. These segments meet only at their endpoints, called **vertices** (plural of **vertex**). The angles inside the polygon are **interior angles**. In a **regular polygon**, all the interior angles are congruent and all of the sides are congruent. When a side of a polygon is extended, it forms an **exterior** angle. An interior and exterior angle at a given vertex are supplementary.

Sum of the Interior Angle Measures in a Polygon	If a polygon has n sides, then $n - 2$ triangles are formed, and the sum of the degree measures of the interior angles of the polygon is $(n - 2)180$.

Examples

a. What is the sum of the measures of the interior angles of a heptagon?

A heptagon has 7 sides, so let $n = 7$.
$(n - 2)180 = (7 - 2)180$
$\qquad\qquad\quad = (5)180$ or 900
The sum of the measures of the interior angles of a heptagon is $900°$.

b. What is the measure of each exterior angle of a regular pentagon?

A pentagon has 5 sides, so the sum of the measures of the interior angles is $(5 - 2)(180)$ or $540°$. Thus, each interior angle measures $540 \div 5$ or $108°$. Each exterior angle measures $180 - 108$ or $72°$.

Try These Together

1. Find the sum of the measures of the interior angles of a triangle.
   HINT: Use the formula and replace n with 3.

2. What is the measure of each exterior angle of a regular hexagon?
   HINT: A hexagon has 6 sides.

Practice

Find the sum of the measures of the interior angles of each polygon.

3. octagon   4. 12-gon   5. 18-gon   6. 30-gon

Find the measure of each exterior angle and each interior angle of each regular polygon. Round to the nearest tenth if necessary.

7. regular triangle   8. regular quadrilateral   9. regular heptagon
10. regular octagon   11. 15-gon   12. 25-gon

Find the perimeter of each regular polygon.

13. regular hexagon with sides 8 cm long   14. regular 17-gon with sides 3 mm long

15. **Standardized Test Practice** What is the perimeter of a regular octagon if the length of one side is 12 inches?

   A 144 in.   B 96 in.   C 84 in.   D 72 in.

Answers: 1. 180° 2. 60° 3. 1080° 4. 1800° 5. 2880° 6. 5040° 7. 120°; 60° 8. 90°; 90° 9. 51.4°; 128.6° 10. 45°; 135° 11. 24°; 156° 12. 14.4°; 165.6° 13. 48 cm 14. 51 mm 15. B

© Glencoe/McGraw-Hill   Glencoe Pre-Algebra

NAME _____ DATE _____ PERIOD _____

10-7 Circumference and Area: Circles (Pages 533–538)

A **circle** is the set of all points in a plane that are the same distance from a given point, called the **center**. The distance from the center to any point on the circle is called the **radius**. The distance across the circle through its center is the **diameter**. The **circumference** is the distance around the circle. The ratio of the circumference to the diameter of any circle is always π **(pi)**, a Greek letter that represents the number 3.1415926.... Pi is an irrational number, however, 3.14 and $\frac{22}{7}$ are considered accepted rational approximations for π.

Circumference of a Circle	The circumference of a circle is equal to the diameter of the circle times π, or 2 times the radius times π. $C = \pi d$ or $C = 2\pi r$ (note: $d = 2r$ or $r = \frac{d}{2}$)
Area of a Circle	The area of a circle is equal to π times the radius squared. $A = \pi r^2$

Examples
Find the circumference and area of each circle to the nearest tenth.

a. The radius is 3 cm.

$C = 2\pi r$ Formula for circumference
$C = 2\pi(3)$ Substitute 3 for r.
$C = 18.8$ cm

$A = \pi r^2$ Formula for area
$A = \pi(3^2)$ Substitute 3 for r.
$A = 28.3$ cm²

b. The diameter is 12 in.

$C = \pi d$ Formula for circumference
$C = \pi(12)$ Substitute 12 for d.
$C = 37.7$ in

$A = \pi r^2$ Formula for area
$A = \pi(6^2)$ $r = \frac{d}{2}$
$A = 113.1$ in²

Practice
Find the circumference and area of each circle to the nearest tenth.

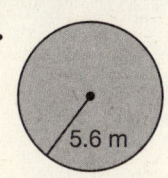

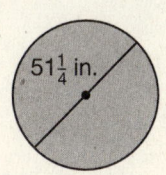

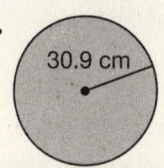

1. (circle with radius 5.6 m)
2. (circle with diameter $51\frac{1}{4}$ in.)
3. (circle with radius 30.9 cm)

4. The diameter is 19 mm.
5. The radius is 25 yd.
6. The radius is 13.8 m.

7. The diameter is 46.2 cm.
8. The radius is $3\frac{1}{4}$ in.
9. The diameter is 6.8 m.

10. **Landscaping** A sprinkler can spray water 10 feet out in all directions. How much area can the sprinkler water?

11. **Standardized Test Practice** What is the area of a half circle whose diameter is 8 m?
    A 25.1 m²   B 50.3 m²   C 100.5 m²   D 201.1 m²

Answers: 1. 35.2 m; 98.5 m² 2. 161 in; 2062.9 in² 3. 194.2 cm; 2999.6 cm² 4. 59.7 mm; 283.5 mm² 5. 157.1 yd; 1963.5 yd² 6. 86.7 m; 598.3 m² 7. 145.1 cm; 1676.4 cm² 8. 20.4 in; 33.2 in² 9. 21.4 m; 36.3 m² 10. 314.2 ft² 11. A

© Glencoe/McGraw-Hill   Glencoe Pre-Algebra

NAME _____ DATE _____ PERIOD _____

10-8 Area: Irregular Figures *(Pages 539–543)*

You have learned the formulas for the area of a triangle, a parallelogram, a trapezoid, and a circle. You can use these formulas to find the area of irregular figures. An irregular figure is a two-dimensional figure that is not one of the previously named shapes. To find the area of an irregular figure, divide the figure into a series of shapes whose area formula you do know. Find the area of each shape. Then, find the sum of the areas of each shape.

Example

Find the area of the figure.

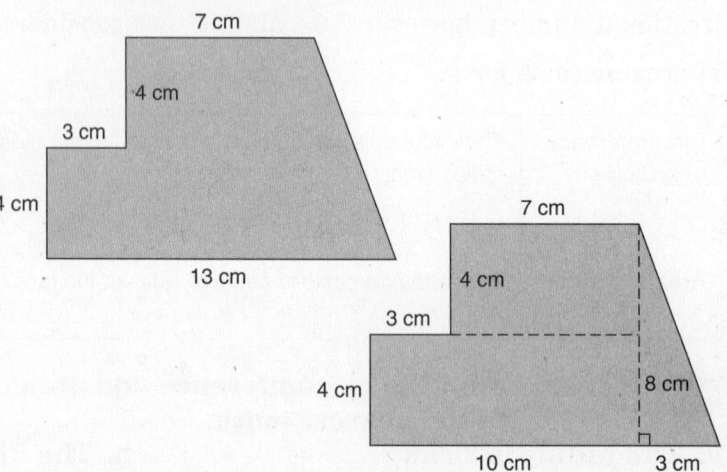

Divide the figure into familiar shapes.
Find the area of each shape.

Rectangle$_1$: $A = \ell w = 7 \cdot 4 = 28$ units^2
Rectangle$_2$: $A = \ell w = 10 \cdot 4 = 40$ units^2
Triangle: $A = \frac{1}{2}bh = 0 \cdot 5 \cdot 3 \cdot 8 = 12$ units^2

Find the sum of the areas.
$28 + 40 + 12 = 80$ units^2

Practice

Find the area of each figure to the nearest tenth, if necessary.

1.

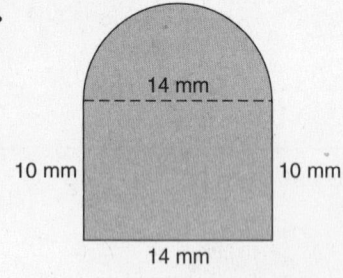

2.

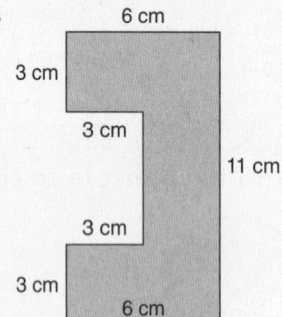

3.

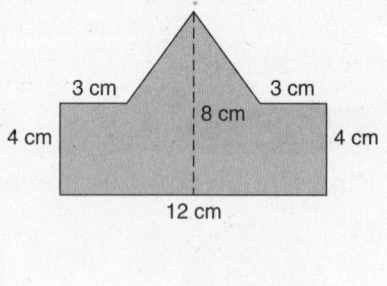

4. **Standardized Test Practice** Which of the following has the same area as the figure shown?

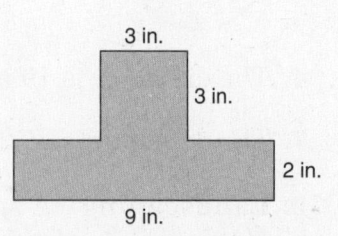

A Square with $s = 5$ in.

B Rectangle with $\ell = 3$ in. and $w = 2.5$ in.

C Triangle with $b = 6$ in. and $h = 9$ in.

D Parallelogram with $b = 4$ in. and $h = 7$ in.

Answers: 1. 217 mm^2 2. 51 cm^2 3. 60 cm^2 4. C

© Glencoe/McGraw-Hill   88   Glencoe Pre-Algebra

10 Chapter Review

Magic Square

Find the value of *x* in each figure. Write each answer in the appropriate square.

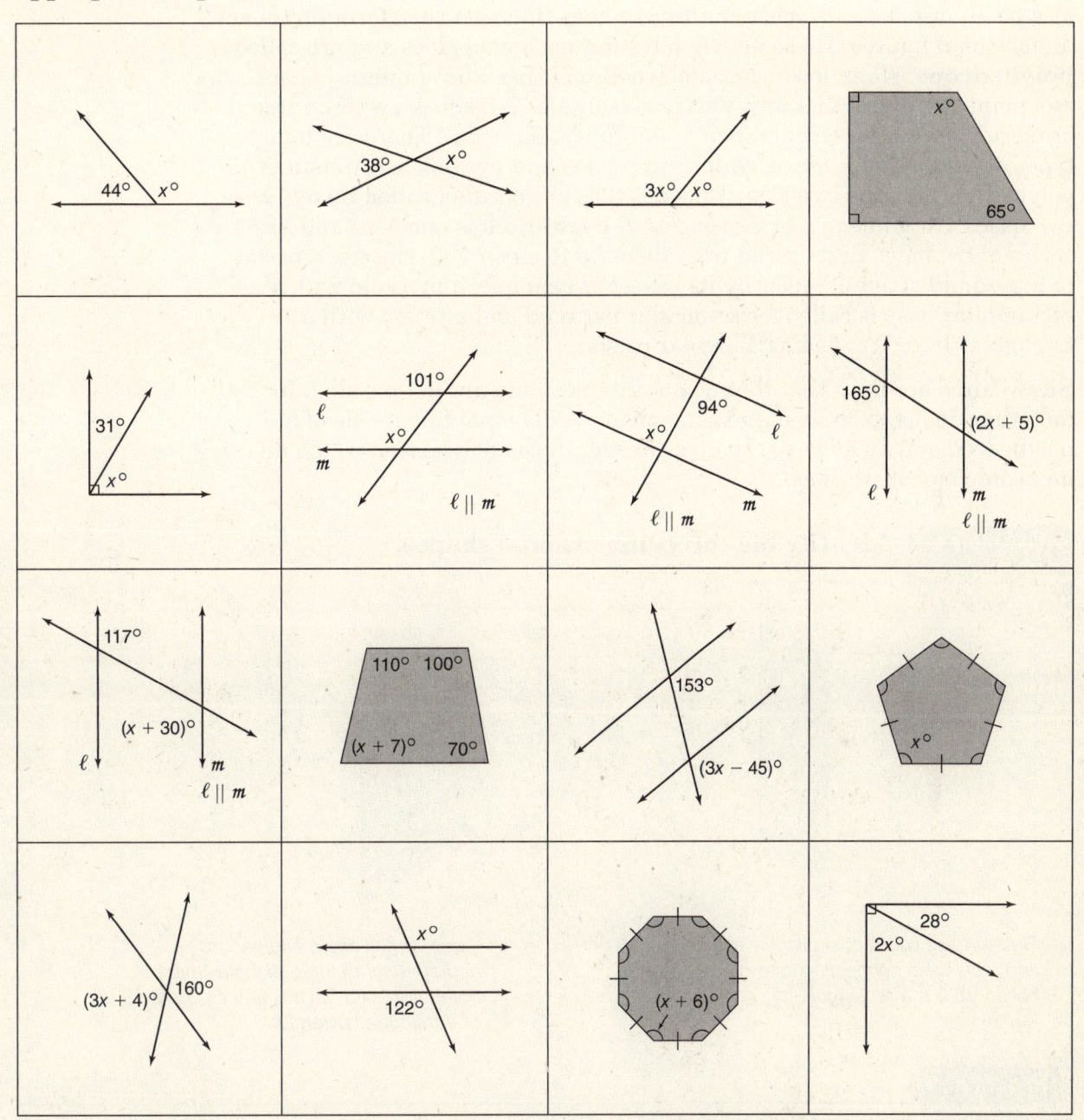

44°, x° (straight line)	38°, x° (vertical angles)	3x°, x° (straight line)	x°, 65° (trapezoid with two right angles)
31°, x° (complementary)	101°, x°, ℓ ∥ m	94°, x°, ℓ ∥ m	165°, (2x + 5)°, ℓ ∥ m
117°, (x + 30)°, ℓ ∥ m	110°, 100°, (x + 7)°, 70°	153°, (3x − 45)°	regular pentagon, x°
(3x + 4)°, 160°	x°, 122°	regular octagon, (x + 6)°	28°, 2x°

The above 4-by-4 square is called a *magic square* because the sum of the answers in each row, column, or main diagonal is the same number. Check to see if you found the correct value of each *x* by finding the sum of each row, column, and main diagonal. Make any needed corrections. What is the correct sum of each row, column, or main diagonal?

Answers are located in the Answer Key.

© Glencoe/McGraw-Hill   89   Glencoe Pre-Algebra

NAME _____ DATE _____ PERIOD _____

11-1 Three-Dimensional Figures *(Pages 556–561)*

A flat surface that contains at least three noncollinear points and extends infinitely in all directions is called a **plane**. Planes can intersect in a line, at a point, or not at all. When multiple planes intersect they form three-dimensional figures. These figures have flat polygonal sides and are called **polyhedrons**. When looking at a polyhedron it is made of **edges**, where two planes intersect in a line, **vertices** (singular is **vertex**), where three or more planes intersect at a point, and **faces**, flat sides. There are many types of polyhedrons, two of which are prisms and pyramids. A **prism** is a polyhedron that has two identical sides that are parallel called **bases**. The two bases are connected by rectangles. A **pyramid** has one base and has a series of triangles that extend from the base to a point. To classify a prism or a pyramid you must identify its base. For example, a pyramid with a rectangular base is called a rectangular pyramid and a prism with a triangular base is called a triangular prism.

Skew lines are lines that that do not intersect and are not parallel. In fact, they do not even lie in the same plane. A diagonal line inside of a polyhedron and an edge on the opposite side of the polyhedron would be an example of skew lines.

Examples   Identify the three-dimensional shapes.

a.

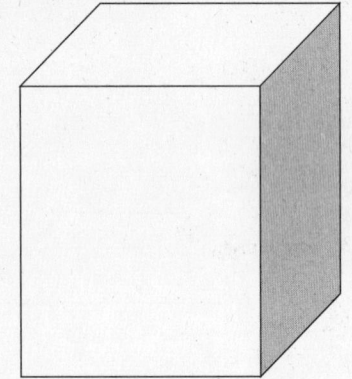

Figure A has two rectangular bases and rectangles connecting its two bases, so it is a rectangular prism.

b.

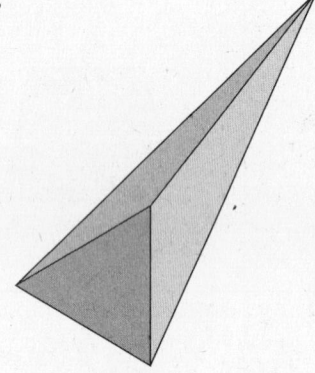

Figure B has one triangular base and consists of three triangles that meet at a point, so it is and example of a triangular pyramid.

Practice
Name the polyhedron.

1.

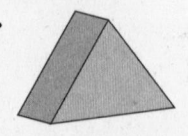

2.

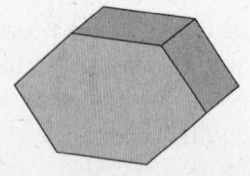

3.

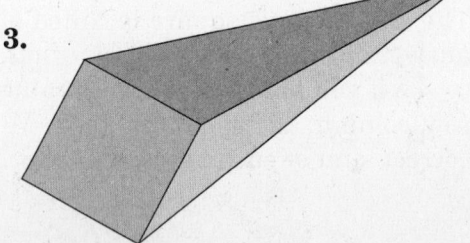

Answers: 1. triangular prism 2. hexagonal prism 3. rectangular pyramid

© Glencoe/McGraw-Hill   Glencoe Pre-Algebra

NAME _____ DATE _____ PERIOD _____

11-2 Volume: Prisms and Cylinders (Pages 563–567)

The amount a container will hold is called its capacity, or **volume**. Volume is often measured in cubic units such as the cubic centimeter (cm³) and the cubic inch (in³).

Volume of a Prism	If a prism has a base area of B square units and a height of h units, then the volume V is B · h cubic units, or V = Bh.
Volume of a Cylinder	If a circular cylinder has a base with a radius of r units and a height of h units, then the volume V is πr²h cubic units, or V = πr²h.

Examples Find the volume of the given figures.

a. a rectangular prism with a length of 3 cm, a width of 4 cm, and a height of 12 cm

V = Bh   Formula for the volume of a prism
V = (ℓw)h   Since the base of the prism is a rectangle, B = ℓw.
V = (3)(4)(12)   Substitute.
V = 144 cm³   Multiply.

b. a circular cylinder with a diameter of 10 in. and a height of 18 in.

V = πr²h   Formula for the volume of a cylinder
V = π(5)²(18)   The diameter is 10 in., so the radius is 5 in.
V = π(25)(18)   Multiply.
V ≈ 1413.7 in³   Simplify.

Try These Together

Find the volume of each solid. Round to the nearest tenth.

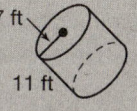

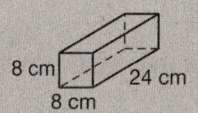

1. 2 m, 1.8 m, 1.2 m, 1.6 m
2. 7 ft, 11 ft
3. 8 cm, 8 cm, 24 cm

HINT: Find the area of the base first, then multiply by the height to get the volume.

Practice

Find the volume of each solid. Round to the nearest tenth.

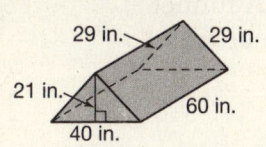

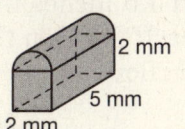

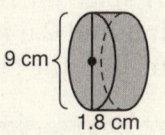

4. 29 in., 29 in., 21 in., 40 in., 60 in.
5. 2 mm, 5 mm, 2 mm
6. 9 cm, 1.8 cm

7. **Landscaping** Nat buys mulch for his flower gardens each fall. How many cubic feet of mulch can he bring home if his truck bed is 5 feet by 8 feet by 2 feet?

8. **Standardized Test Practice** What is the height of a cylindrical prism whose volume is 141.3 cubic meters and whose diameter is 10 meters?

    A 0.45 m   **B** 0.9 m   **C** 1.8 m   **D** 2.25 m

Answers: 1. 1.7 m³ 2. 1693.3 ft³ 3. 1536 cm³ 4. 25,200 in³ 5. 27.9 mm³ 6. 114.5 cm³ 7. 80 ft³ 8. C

© Glencoe/McGraw-Hill   Glencoe Pre-Algebra

NAME _____ DATE _____ PERIOD _____

11-3 Volume: Pyramids and Cones (Pages 568–572)

When you find the volume of a pyramid or cone, you must know the height h. The height is *not* the same as the lateral height, which you learned in an earlier lesson. The height h of a pyramid or cone is the length of a segment from the vertex to the base, perpendicular to the base.

Volume of a Pyramid	If a pyramid has a base of B square units, and a height of h units, then the volume V is $\frac{1}{3} \cdot B \cdot h$ cubic units, or $V = \frac{1}{3}Bh$.
Volume of a Cone	If a cone has a radius of r units and a height of h units, then the volume V is $\frac{1}{3} \cdot \pi \cdot r^2 \cdot h$ cubic units, or $V = \frac{1}{3}\pi r^2 h$.

Examples Find the volume of the given figures.

a. a square pyramid with a base side length of 6 cm and a height of 15 cm

$V = \frac{1}{3}Bh$ Formula for the volume of a pyramid

$V = \frac{1}{3}s^2h$ Replace B with s^2.

$V = \frac{1}{3}(6)^2(15)$ or 180 cm³

b. a cone with a radius of 3 in. and a height of 8 in.

$V = \frac{1}{3}\pi r^2 h$ Formula for the volume of a cone

$V = \frac{1}{3}\pi(3)^2(8)$ $r = 3$ and $h = 8$

$V = \frac{1}{3}\pi(9)(8)$ or about 75.4 in³

Practice

Find the volume of each solid. Round to the nearest tenth.

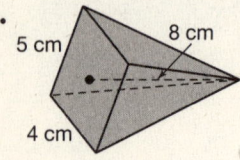

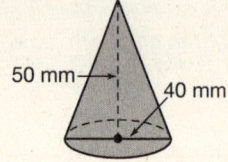

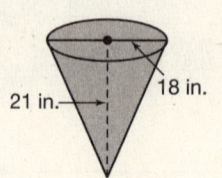

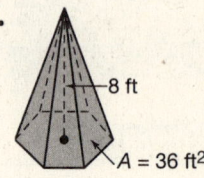

1. 5 cm, 8 cm, 4 cm
2. 50 mm, 40 mm
3. 21 in., 18 in.
4. 8 ft, $A = 36$ ft²

5. **Cooking** A spice jar is 3 inches tall and 1.5 inches in diameter. A funnel is 2 inches tall and 2.5 inches in diameter. If Hayden fills the funnel with pepper to put into the spice jar, will it overflow?

6. **Standardized Test Practice** A square pyramid is 6 feet tall and with the sides of the base 8 feet long. What is the volume of the pyramid?
   A 96 ft³   B 128 ft³   C 192 ft³   D 384 ft³

Answers: 1. 53.3 cm³ 2. 20,944.0 mm³ 3. 1781.3 in³ 4. 96 ft³ 5. No; jar volume ≈ 5.3 in³, funnel volume ≈ 3.3 in³ 6. B

© Glencoe/McGraw-Hill   92   Glencoe Pre-Algebra

NAME _____ DATE _____ PERIOD ____

11-4 Surface Area: Prisms and Cylinders (Pages 573–577)

In geometry, a solid like a cardboard box is called a **prism**. A prism is a solid figure that has two parallel, congruent sides, called **bases**. A prism is named by the shape of its bases. For example, a prism with rectangular-shaped bases is a **rectangular prism**. A prism with triangular-shaped bases is a **triangular prism**. A **cylinder** is a geometric solid whose bases are parallel, congruent circles. The **surface area** of a geometric solid is the sum of the areas of each side or **face** of the solid. If you open up or unfold a prism, the result is a **net**. Nets help you identify all the faces of a prism.

A triangular prism has five faces.

Examples Find the surface area of the given geometric solids.

a. a box measuring 6 in. × 8 in. × 12 in.

Find the surface area of the faces. Use the formula $A = \ell w$. Multiply each area by 2 because there are two faces with each area.

Front and Back: 6 × 8 = 48 (each)
Top and Bottom: 12 × 8 = 96 (each)
Two Sides: 6 × 12 = 72 (each)
Total: 2(48) + 2(96) + 2(72) = 432 in²

b. a cylinder with a radius of 10 cm and a height of 24 cm

The surface area of a cylinder equals the area of the two circular bases, $2\pi r^2$, plus the area of the curved surface. If you make a net of a cylinder, you see that the curved surface is really a rectangle with a height that is equal to the height h of the cylinder and a length that is equal to the circumference of the circular bases, $2\pi r$.

Surface area = $2\pi r^2 + h \cdot 2\pi r$
Surface area = $2\pi(100) + 48\pi(10)$
Surface area ≈ 628.3 + 1508.0
Surface area ≈ 2136.3 cm²

Practice

Find the surface area of each solid. Round to the nearest tenth.

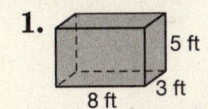

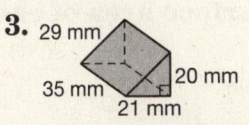

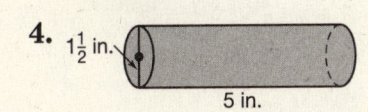

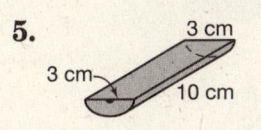

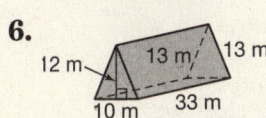

1. 8 ft × 3 ft × 5 ft
2. cylinder: 4.7 m, 12 m
3. triangular prism: 29 mm, 35 mm, 21 mm, 20 mm
4. cylinder: $1\frac{1}{2}$ in., 5 in.
5. cylinder: 3 cm, 3 cm, 10 cm
6. triangular prism: 12 m, 13 m, 13 m, 10 m, 33 m

7. **Pets** A pet store sells nylon tunnels for dog agility courses. If a tunnel is 6 feet long and $1\frac{1}{2}$ feet in diameter, how many square feet of nylon is used?

8. **Standardized Test Practice** The height of a cylinder is 10 meters and its diameter is 4 meters. What is its surface area?

   A 75.4 m²   B 138.2 m²   C 150.8 m²   D 351.9 m²

Answers: 1. 158 ft² 2. 211.9 m² 3. 2,870 mm² 4. 27.1 in² 5. 54.2 cm² 6. 1,308 m² 7. about 28.3 ft² 8. C

© Glencoe/McGraw-Hill   Glencoe Pre-Algebra

NAME _____ DATE _____ PERIOD _____

11-5 Surface Area: Pyramids and Cones (Pages 578–582)

A **pyramid** is a solid figure that has a polygon for a base and triangles for sides, or *lateral* faces. Pyramids have just one base. The lateral faces intersect at a point called the **vertex**. Pyramids are named for the shapes of their bases. For example, a **triangular pyramid** has a triangle for a base. A **square pyramid** has a square for a base. The **slant height** of a pyramid is the altitude of any of the lateral faces of the pyramid. To find the surface area of a pyramid, you must find the area of the base and the area of each lateral face. The area of the **lateral surface** of a pyramid is the area of the lateral faces (not including the base). A **circular cone** is another solid figure and is shaped like some ice cream cones. Circular cones have a circle for their base.

Surface Area of a Circular Cone	The surface area of a cone is equal to the area of the base, plus the lateral area of the cone. The surface area of the base is equal to πr^2. The lateral area is equal to $\pi r \ell$, where ℓ is the slant height of the cone. So, the surface area of the cone, SA, is equal to $\pi r^2 + \pi r \ell$.

Examples
Find the surface area of the given geometric solids.

a. a square pyramid with a base that is 20 m on each side and a slant height of 40 m

Find the surface area of the base and the lateral faces.
Base:
$A = s^2$ or $(20)^2$
$A = 400$
$SA = 400 + 4(400)$
$SA = 2000$ m^2

Each triangular side:
$A = \frac{1}{2}bh$ or $\frac{1}{2}(20)(40)$
$A = 400$
Area of the base plus area of the four lateral sides.

b. a cone with a radius of 4 cm and a slant height of 12 cm

Use the formula.
$SA = \pi r^2 + \pi r \ell$
$SA \approx \pi(4)^2 + \pi(4)(12)$
$SA \approx 50.3 + 150.8$
$SA \approx 201.1$ cm^2

Practice
Find the surface area of each solid. Round to the nearest tenth.

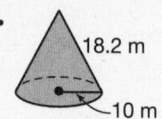

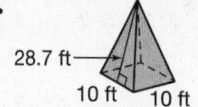

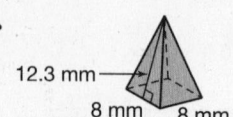

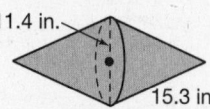

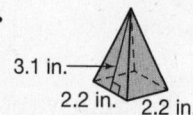

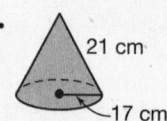

1. cone: 18.2 m, 10 m
2. pyramid: 28.7 ft, 10 ft, 10 ft
3. pyramid: 12.3 mm, 8 mm, 8 mm
4. 11.4 in., 15.3 in.
5. pyramid: 3.1 in., 2.2 in., 2.2 in.
6. cone: 21 cm, 17 cm

7. **Standardized Test Practice** What is the surface area of a square pyramid where the length of each side of the base is 10 meters and the slant height is also 10 meters?

A 300 m^2 B 400 m^2 C 500 m^2 D 1000 m^2

Answers: 1. 885.9 m^2 2. 674 ft^2 3. 260.8 mm^2 4. 548.0 in^2 5. 18.5 in^2 6. 2029.5 cm^2 7. A

© Glencoe/McGraw-Hill   94   Glencoe Pre-Algebra

NAME _____ DATE _____ PERIOD _____

11-6 Similar Solids *(Pages 584–588)*

A pair of three-dimensional figures is classified as **similar solids** when they are the same shapes and their corresponding measurements are proportional. The ratio that compares the measurements of two similar solids is called the **scale factor**.

> Given two similar solids Figure A and Figure B:
> - The scale factor of corresponding sides of Figure A to Figure B is $\frac{a}{b}$.
> - The ratio of the surface area of Figure A to the surface area of Figure B is $\frac{a^2}{b^2}$.
> - The ratio of the volume of Figure A to the volume of Figure B is $\frac{a^3}{b^3}$.

Examples

The square prisms to the right are similar. Find the scale factor, the ratio of their surface areas, and the ration of their volumes.

The scale factor is

$$\frac{a}{b} = \frac{20}{5} = 4$$

The ratio of the surface areas is

$$\frac{a^2}{b^2} = \frac{4^2}{1^2} = 16$$

The ratio of the volumes is

$$\frac{a^3}{b^3} = \frac{4^3}{1^3} = 64$$

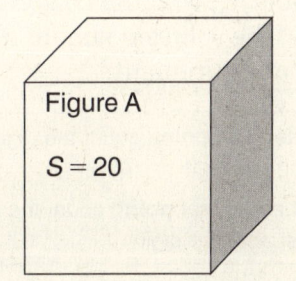

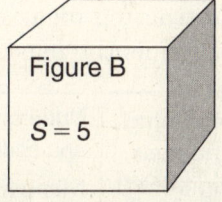

Figure A $S = 20$

Figure B $S = 5$

Practice

Triangular Prism X and triangular Prism Y are similar. The scale factor of Prism X to Prism Y is $\frac{3}{4}$. Use this information for problem 1–4.

1. If the length of a side of Prism X is 9 feet, what is the length of the corresponding side of Prism Y?

2. If Prism X has a surface area of 88.8 feet^2, what is the surface area of Prism Y?

3. If the volume of Prism X is 35.1 feet^3, what is the volume of Prism Y?

4. **Standardized Test Practice** The height of the triangular base of Prism Y is 3.5 feet. Find the height of the triangular base of Prism X.

   A 4.7 feet   B 6.2 feet   C 8.3 feet   D 2.6 feet

Answers: 1. 12 feet 2. 157.9 feet^2 3. 83.2 feet^3 4. D

© Glencoe/McGraw-Hill   95   Glencoe Pre-Algebra

11-7 Precision and Significant Digits *(Pages 590–594)*

The smallest unit of measure used for a particular measurement, known as the precision unit, dictates the **precision**. When measuring an object you can round to the nearest precision unit, but a more precise method is to include all known digits plus an estimated unit. These digits, the known and the estimated, are called **significant digits**. Let's say you are measuring the length of your calculator with a standard ruler. The precision unit of the ruler is $\frac{1}{16}$ inch. You can measure to the nearest $\frac{1}{16}$ inch, we'll say the calculator was $7\frac{7}{8}$ inches or 7.875 inches. This is a rounded version of the measurement to the precision unit; however, we could be more precise by using estimation. Let's say that when closely reviewing the measurement we find the calculator was actually slightly bigger than $7\frac{7}{8}$, or 7.875 inches. In fact, the calculator was almost half way between $7\frac{7}{8}$ and $7\frac{15}{16}$. Therefore, we could estimate the calculator to be $7\frac{14.5}{16} = 7\frac{29}{32}$ or 7.90625 inches. This more precise measurement is an example of significant digits. The number 7.90625 has 6 significant digits. When adding or subtracting measurements, the solution should always have the same precision as the least precise measurement.

Determining the number of significant digits	**Numbers with a decimal point**: count the digits from left to right starting with the first nonzero digit and ending with the last digit
	Numbers without a decimal point: count the digits from left to right starting with the first digit and ending with the last nonzero digit

Examples Find the number of significant digits.

a. 3.43
3

b. 0.005
1

c. 240
2

d. 6.70
3

Practice

Compute using significant digits.

1. Find the perimeter of a rectangle with length 10.255 cm and with width 7.1 cm.

2. Find the perimeter of a triangle with sides of length 3.1 m, 12.02 m, and 7.223 m.

3. What is the area of a parallelogram with length 17.25 mm and width 5.065 mm?

4. **Standardized Test Practice** How many significant digits are in the number 0.00016?
   A 6   B 5   C 2   D 1

Answers: 1. 34.7 2. 22.3 3. 87.37 4. C

© Glencoe/McGraw-Hill   96   Glencoe Pre-Algebra

NAME _____ DATE _____ PERIOD _____

11 Chapter Review

Robots

This robot is made of common three-dimensional figures.

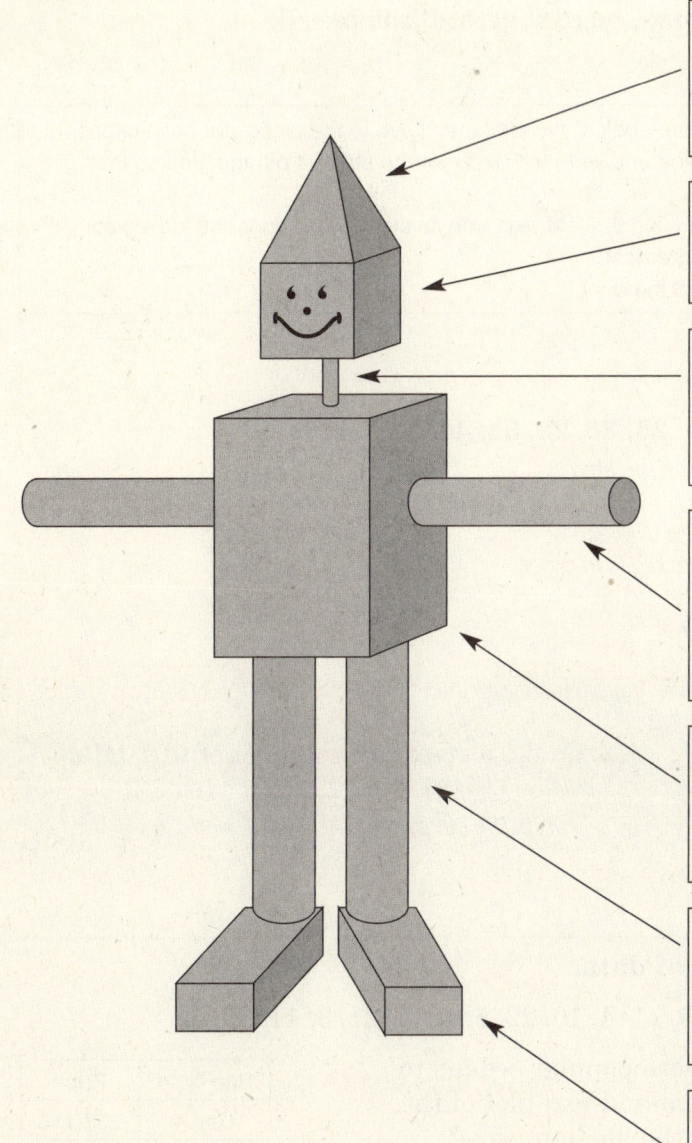

The hat is a square pyramid. Each side of the base is 6 inches long, and the height of the pyramid is 8 inches. What is the volume of the hat?

The head is a cube. Each side of the cube is 6 inches long. Find the volume of the head.

The neck is a cylinder. The radius of the base is 1 inch and the height of the cylinder is 3 inches. What is the volume of the neck to the nearest whole number.

Cylinders are used for the arms. The diameter of each arm is 3 inches and the length of each arm is 15 inches. Find the volume of one arm to the nearest whole number.

The torso is a rectangular prism. The dimensions of the body are 10 inches by 10 inches by 15 inches. What is the volume of the torso?

Cylinders are used for the legs. Each leg is 4 inches in diameter and 18 inches long. Find the volume of one leg to the nearest whole number.

Rectangular prisms are used for feet. Each foot is 5 feet by 3 feet by 6 feet. What is the volume of each foot?

What is the total volume of the robot?

Answers are located in the Answer Key.

© Glencoe/McGraw-Hill   Glencoe Pre-Algebra

NAME _____ DATE _____ PERIOD _____

12-1 Stem-and-Leaf Plots (Pages 606–611)

One way to organize a set of data and present it in a way that is easy to read is to construct a **stem-and-leaf plot**. Use the greatest place value common to all the data values for the **stems**. The next greatest place value forms the **leaves**.

Making a Stem-and-Leaf Plot	1. Find the least and greatest value. Look at the digit they have in the place you have chosen for the stems. Draw a vertical line and write the digits for the stems from the least to the greatest value. 2. Put the leaves on the plot by pairing the leaf digit with its stem. Rearrange the leaves so they are ordered from least to greatest. 3. Include an explanation or key of the data.

Example

Make a stem-and-leaf plot of this data: 25, 36, 22, 34, 44, 33, 26, 48

The greatest place value is the tens place, so that will be the stems. The ones place will be the leaves.

1. The least value is 22 and the greatest is 48. This data uses stems of 2, 3, and 4. Draw a vertical line and write the stem digits in order.
2. Put on the leaves by pairing each value.
3. Include an explanation. Since 4|8 represents 48, 4|8 = 48.

Step 1:
```
2|
3|
4|
```

Step 2:
```
2 | 2 5 6
3 | 3 4 6
4 | 4 8
```
4|8 = 48

Try These Together

1. Make a stem-and-leaf plot of this data: 12, 43, 42, 18, 27, 33, 12, 22.
   HINT: The stems are 1, 2, 3, and 4.

2. Make a stem-and-leaf plot of this data: 105, 115, 91, 109, 120, 81, 114, 119.
   HINT: The stems are 8, 9, 10, 11, and 12.

Practice

Make a stem-and-leaf plot of each set of data.

3. 5.3, 5.1, 6.1, 6.3, 5.7, 8.9, 6.8, 8.1, 9, 5.9

4. 10, 22, 5, 18, 7, 21, 3, 11, 30, 15

5. **Automobiles** Round the prices of these popular sedans to the nearest hundred. Then make a stem-and-leaf plot of the prices. (Use 36|4 = $36,400.) What is the median price? Explain whether you think the table or the stem-and-leaf plot is a better representation of the data.

Car Type	Price
Car A	$33,158
Car B	$30,710
Car C	$30,855
Car D	$31,600
Car E	$29,207
Car F	$28,420
Car G	$30,535

6. **Standardized Test Practice** What is the median of grades in Mrs. Jones' class?
   A 85   B 86   C 87   D 88

```
7 | 3 6 7
8 | 3 5 6 7 8 8
9 | 2 5 5 6 8 9
```
9|2 = 92

Answers: 1–5. See Answer Key. 6. D

© Glencoe/McGraw-Hill   98   Glencoe Pre-Algebra

NAME _____ DATE _____ PERIOD _____

12-2 Measures of Variation (Pages 612–616)

The **range** of a set of numbers is the difference between the least and greatest number in the set. In a large set of data, it is helpful to separate the data into four equal parts called **quartiles**. The *median* of a set of data separates the data in half. The median of the lower half of a set of data is the **lower quartile (LQ)**. The median of the upper half of the data is called the **upper quartile (UQ)**.

Finding the Interquartile Range	The **interquartile range** is the range of the middle half of a set of numbers. Interquartile range = UQ − LQ

Example

Find the range, median, UQ, LQ, and interquartile range: 5, 7, 3, 9, 6, 9, 4, 6, 7

First list the data in order from least to greatest: 3, 4, 5, 6, 6, 7, 7, 9, 9.
The range is 9 − 3 or 6.
Next find the median, UQ, and LQ.

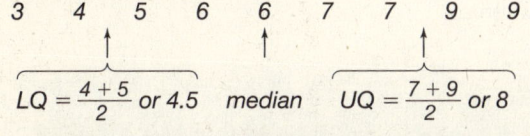

3   4   5   6   6   7   7   9   9

$LQ = \frac{4+5}{2}$ or 4.5   median $UQ = \frac{7+9}{2}$ or 8   The interquartile range is 8 − 4.5 or 3.5.

Try These Together

Find the range, median, upper and lower quartiles, and the interquartile range for each set of data.

1. 20, 90, 80, 70, 50, 40, 90
2. 67°, 52°, 60°, 58°, 62°

HINT: First arrange the data in order from least to greatest.

Practice

Find the range, median, upper and lower quartiles, and the interquartile range for each set of data.

3. 30, 54, 42, 45, 61, 44, 62, 57, 59, 53

4. 101, 128, 124, 129, 120, 108, 102, 118, 127, 123, 116

5. 78, 84, 100, 69, 70, 75, 87, 85, 97, 89

6. **Measurement** The following list gives the heights in inches of a group of people. Find the range and the interquartile range for the data.
   48, 55, 50, 52, 49, 55, 60, 61, 62, 56, 53

7. **Standardized Test Practice** What is the lower quartile of the set of data?
   9, 10, 7, 4, 20, 17, 12, 8, 5, 16, 21, 0, 8, 13
   A 8   **B** 7   **C** 6   **D** 5

Answers: 1. 70; 70; 90, 40, 50 2. 15; 60; 64.5, 55; 9.5 3. 32; 53.5; 59, 44; 15 4. 28; 120; 127, 108; 19 5. 31; 84.5; 89, 75; 14 6. 14; 10 7. B

© Glencoe/McGraw-Hill   99   Glencoe Pre-Algebra

NAME _____ DATE _____ PERIOD _____

12-3 Box-and-Whisker Plots *(Pages 617–621)*

One way to display data is with a **box-and-whisker plot**. This kind of plot summarizes data using the median, the upper and lower quartiles, and the highest and lowest, or extreme, values.

Drawing a Box-and-Whisker Plot	1. Draw a number line for the range of the values. Above the number line, mark points for the extreme, median, and quartile values. 2. Draw a box that contains the quartile values. Draw a vertical line through the median value. Then extend the whiskers from each quartile to the extreme data points.

Example

Draw a box-and-whisker plot for this data: 5, 7, 3, 9, 6, 9, 4, 6, 7

1. Arrange the data in order from least to greatest (3, 4, 5, 6, 6, 7, 7, 9, 9) and find the extreme (3 and 9), the median (6), the upper quartile (8) and the lower quartile (4.5). Draw a number line and mark these points.
2. Draw a box that contains the quartile values and a vertical line through the median. Then extend the whiskers from each quartile to the extremes.

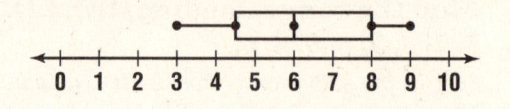

Try These Together

1. What is the median for the plot shown in PRACTICE below?
2. What is the upper quartile for the plot shown in PRACTICE below?

HINT: The median is the point that divides the data in half. The upper quartile is the middle of the upper half.

Practice

Use the stem-and-leaf plot at the right to answer each question.

```
5 | 0
6 | 1 3
7 | 0 5
8 | 0 3 5 9
9 | 1 2 3 5 9
5|0 = 50
```

3. What is the lower quartile?
4. Make a box-and-whisker plot of the data.
5. What is the interquartile range?
6. What are the extremes?
7. To the nearest 25%, what percent of the data is represented by each whisker?
8. Why isn't the median in the middle of the box?
9. What percent of data does the box represent?
10. To the nearest 25%, what percent of data is above the upper quartile?

11. **Standardized Test Practice** What is the best way to display the table of world population data?

Year	1	1000	1250	1500	1750	1800	1850	1900	1950
Billion	0.30	0.31	0.40	0.50	0.79	0.98	1.26	1.65	2.52

A circle graph **B** stem-and-leaf plot **C** box-and-whisker plot **D** line graph

Answers: 1. 84 2. 92 3. 70 4. See Answer Key. 5. 22 6. 50, 99 7. 25% 8. The median isn't necessarily the mean (average) of the upper and lower quartiles; it is the midpoint of the data between the upper and lower quartiles. 9. 50% 10. 25% 11. D

© Glencoe/McGraw-Hill   100   Glencoe Pre-Algebra

NAME _____ DATE _____ PERIOD _____

12-4 Histograms *(Pages 623–628)*

A **histogram** is a graph that displays data. Like a bar graph, a histogram uses bars to represent data. The bars in a histogram do not have any gaps between them. In order to construct a histogram, you must have data that is divided into intervals. The number of elements that fall into an interval determines the height of the corresponding bar on a histogram.

Example

Data has been collected on the number of each test score for Mr. Brown's students. Using the data in the table, construct a histogram of the data.

Score	Frequency
0%–19%	6
20%–39%	5
40%–59%	17
60%–79%	53
80%–100%	41

Begin by drawing and labeling a vertical and a horizontal axis. The horizontal axis should show the intervals. For each interval, draw a bar whose height is the frequency.

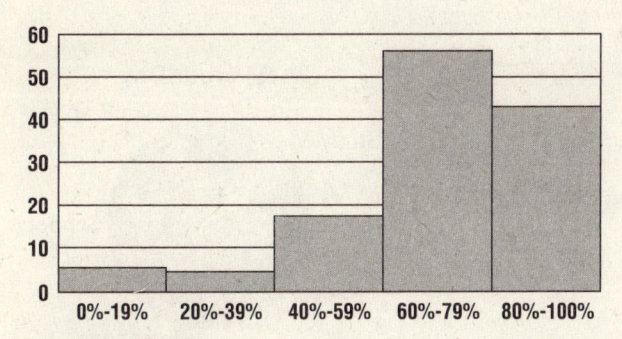

Practice

Use the information from the example to answer the following questions.

1. Which interval has the greatest number of students?
2. Which interval has the least number of students?
3. How many students scored 59% or lower?
4. How many students scored 40% or above?
5. **Standardized Test Practice** Select the answer choice, which represents a true statement, based upon the data in the histogram.

   A More students scored below 60% than above.
   B Mr. Browns test was 65 questions.
   C The second largest interval was 80%–100%.

Answers: 1. 60%–79% 2. 20%–39% 3. 28 4. 111 5. C

© Glencoe/McGraw-Hill   Glencoe Pre-Algebra

NAME _____ DATE _____ PERIOD _____

12-5 Misleading Statistics *(Pages 630–633)*

The same data can be used to support different points of view depending on how that data is displayed.

Looking for Misleading Graphs	Here are some things to check as you decide if a graph is misleading. • Is one of the axes extended or shortened compared to the other? • Are there misleading breaks in an axis? • Are all the parts of the graph labeled clearly? • Does the axis include zero if necessary? • If statistics are compared, do they all use the same measure of central tendency, or does one use the mean and another the median?

Examples

a. What words do you need to put on your graphs?

Graphs need a title and labels on the scales for each axis.

b. What do you check on the scales and the axis when you look for a misleading graph?

Make sure the axis includes 0 if it applies. Check that the distance between the units is uniform. Is the scale chosen to minimize or emphasize change?

Practice

A student made the table below and used it to make the bar graph and circle graph to the right of it.

A 24-Hour Day	
Activity	Hours
Sleep	8
Studying	2
TV	2
Swim Practice	2
School	8
Telephone	1

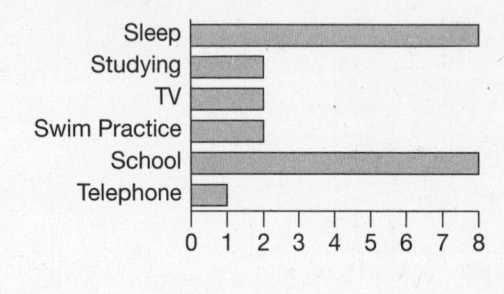

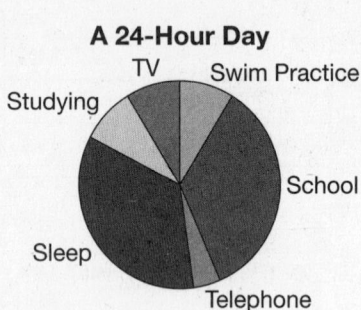

1. What is wrong with the data in the table?
2. What is missing on the bar graph? *(HINT:* Interpret the meaning of the School bar.)
3. What is missing in the circle graph?
4. Compare the visual effects of the bar graph versus the circle graph.
5. **Standardized Test Practice** Generally, the best measure of central tendency is—
   A the mode.
   B the mean.
   C the median.
   D dependent on the data.

Answers: 1. The hours don't add up to 24. 2. title; unit of measure for the x-axis 3. numerical data 4. Answers will vary. 5. D

© Glencoe/McGraw-Hill 102 Glencoe Pre-Algebra

NAME _____ DATE _____ PERIOD _____

12-6 Counting Outcomes (Pages 635–639)

You can use a **tree diagram** or the **Fundamental Counting Principle** to count **outcomes**, the number of possible ways an event can occur.

Fundamental Counting Principle	If an event M can occur in m ways and is followed by event N that can occur in n ways, then the event M followed by event N can occur in $m \cdot n$ ways.

Examples

How many lunches can you choose from 3 different drinks and 4 different sandwiches?

Letter the different sandwiches A, B, C, and D.
A tree diagram shows 12 as the number of outcomes.
You could also use the Fundamental Counting Principle.

number of types of drinks × number of types of sandwiches = number of possible outcomes

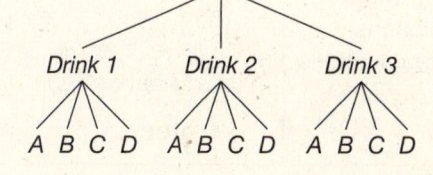

3 × 4 = 12   There are 12 possible outcomes.

Try These Together

1. Draw a tree diagram to find the number of outcomes when a coin is tossed twice.

2. A six-sided number cube is rolled twice. How many possible outcomes are there?

Practice

Draw a tree diagram to find the number of outcomes for each situation.

3. A six-sided number cube is rolled and then a dime is tossed.

4. Julie can either catch the bus or walk to school in the mornings. In the afternoons, she has a choice of catching a ride with a friend, taking the bus, or walking home. How many different ways can Julie get to and from school?

5. **Fast Food** A fast-food restaurant makes specialty burritos. The tortillas come in the sizes of regular, monster, and super and in flavors of wheat, flour, cayenne, and spinach. How many different combinations of size and flavor of tortilla can you order for a burrito?

6. **Standardized Test Practice** Using two six-sided number cubes, what is the probability of rolling two 1s?

A $\frac{1}{36}$ B $\frac{1}{18}$ C $\frac{1}{12}$ D $\frac{1}{6}$

Answers: 1. 4 outcomes; see Answer Key for diagram. 2. 36 outcomes 3–4. See Answer Key for diagrams. 3. 12 outcomes 4. 6 ways 5. 12 combinations 6. A

© Glencoe/McGraw-Hill   Glencoe Pre-Algebra

NAME _____ DATE _____ PERIOD _____

12-7 Permutations and Combinations (Pages 641–645)

An arrangement in which order is important is called a **permutation**. Arrangements or listings where the order is not important are called **combinations**. Working with these arrangements, you will use **factorial** notation. The symbol 5!, or 5 factorial, means $5 \cdot 4 \cdot 3 \cdot 2 \cdot 1$. The expression $n!$ means the product of all counting numbers beginning with n and counting backwards to 1. The definition of 0! is 1.

Working with Permutations and Combinations	• The symbol $P(7, 3)$ means the number of permutations of 7 things taken 3 at a time. To find $P(7, 3)$, multiply the number of choices for the 1st, 2nd, and 3rd positions. $P(7, 3) = 7 \cdot 6 \cdot 5$ or 210 • The symbol $C(7, 3)$ means the number of combinations of 7 things taken 3 at a time. To find $C(7, 3)$, divide $P(7, 3)$ by 3!, which is the number of ways of arranging 3 things in different orders. $C(7, 3) = \frac{P(7, 3)}{3!} = \frac{7 \cdot 6 \cdot 5}{3 \cdot 2 \cdot 1}$ or 35

Examples

a. Find $P(5, 3)$.
$P(5, 3) = 5 \cdot 4 \cdot 3$ or 60

b. Find $C(5, 3)$.
First find the value of P(5, 3). From Example A, you know that P(5, 3) is 60. Divide 60 by 3!. This is $\frac{60}{6}$ or 10.

c. Fred plans to buy 4 tropical fish from a tank at a pet shop. Does this situation represent a permutation or a combination? Explain.
This situation represents a combination. The only thing that matters is which fish he selects. The order in which he selects them is irrelevant.

Practice

Tell whether each situation represents a permutation or combination.

1. a stack of 18 tests
2. two flavors of ice cream out of 31 flavors
3. 1st-, 2nd-, and 3rd-place winners
4. 20 students in a single file line

How many ways can the letters of each word be arranged?

5. RANGE
6. QUARTILE
7. MEDIAN

Find each value.

8. $P(5, 2)$
9. $P(10, 3)$
10. $7!$
11. $9!$
12. $C(7, 2)$
13. $C(12, 3)$
14. $\frac{5!2!}{3!}$
15. $\frac{8!4!}{7!3!}$

16. **Standardized Test Practice** If there are 40 clarinet players competing for places in the district band, how many ways can the 1st and 2nd chairs be filled?

   A 40!   **B** $40 \cdot 39$ **C** $\frac{40 \cdot 39}{2!}$ **D** 2

Answers: 1. permutation 2. combination 3. permutation 4. permutation 5. 120 ways 6. 40,320 ways 7. 720 ways 8. 20 9. 720 10. 5040 11. 362,880 12. 21 13. 220 14. 40 15. 32 16. B

NAME _____ DATE _____ PERIOD _____

12-8 Odds (Pages 646–649)

One way to describe the chance of an event's occurring is by using **odds**.

Finding the Odds	• The odds in favor of an outcome is the ratio of the number of ways the outcome can occur to the number of ways the outcome cannot occur. Odds in favor = number of successes : number of failures • The odds against an outcome is the ratio of the number of ways the outcome cannot occur to the number of ways the outcome can occur. Odds against = number of failures : number of successes

Examples

a. Find the odds of getting a 5 when you roll an eight-sided number cube.

There is only 1 successful outcome: 5. There are 7 failures. The odds are 1:7.

b. Find the odds against getting an even number when you roll an eight-sided number cube.

There are 4 failures and 4 successes, so the odds against are 4:4 or 1:1.

Try These Together

1. Find the odds of rolling a 3 with a six-sided number cube.
2. Find the odds of rolling an odd number with a six-sided number cube.

HINT: Find the number of successes divided by the number of failures.

Practice

Find the odds of each outcome if a six-sided number cube is rolled.

3. the number 4 or 5
4. the number 1, 2, or 3
5. a prime number
6. a factor of 12
7. a multiple of 3
8. a number less than 5
9. a number greater than 6
10. not a 6
11. not a 1, 2, 3, 4, 5, or 6
12. a factor of 10

A bag contains 9 red marbles, 2 blue marbles, 3 black marbles and 1 green marble. Find the odds of drawing each outcome.

13. a green marble
14. a red marble
15. a blue marble
16. a black marble
17. not a black marble
18. a green or red marble

19. **Technology** Adela has noticed that the time of day makes a difference when she is trying to get connected to the Internet. At 4 P.M., she is able to get connected right away 8 times out of 10. What are the odds of getting connected right away at 4 P.M.?

20. **Standardized Test Practice** What are the odds of getting a head when you toss a penny?

A 1:2 B 2:1 C 0:1 D 1:1

Answers: 1. 1:5 2. 1:1 3. 1:2 4. 1:1 5. 1:1 6. 5:1 7. 1:2 8. 2:1 9. 0:6 10. 5:1 11. 0:6 12. 1:1 13. 1:14 14. 3:2 15. 2:13 16. 1:4 17. 4:1 18. 2:1 19. 4:1 20. D

© Glencoe/McGraw-Hill   Glencoe Pre-Algebra

NAME _____ DATE _____ PERIOD _____

12-9 Probability of Compound Events

(Pages 650–655)

Events are **independent** when the outcome of one event does not influence the outcome of a second event. When the outcome of one event affects the outcome of a second event, the events are **dependent**.

When two events cannot happen at the same time, they are **mutually exclusive**.

Finding Probability	• To find the probability of two independent events both occurring, multiply the probability of the first event by the probability of the second event. $P(A \text{ and } B) = P(A) \cdot P(B)$ • To find the probability of two dependent events both occurring, multiply the probability of A and the probability of B after A occurs. $P(A \text{ and } B) = P(A) \cdot P(B \text{ following } A)$ • To find the probability of one or the other of two *mutually exclusive* events, add the probability of the first event to the probability of the second event. $P(A \text{ or } B) = P(A) + P(B)$

Examples

a. Find the probability of tossing two number cubes and getting a 3 on each one.

These events are independent.

$P(3) \cdot P(3) = \frac{1}{6} \cdot \frac{1}{6}$ or $\frac{1}{36}$

The probability is $\frac{1}{36}$.

b. A box contains a nickel, a penny, and a dime. Find the probability of choosing first a dime and then, without replacing the dime, choosing a penny.

These events are dependent. The first probability is $\frac{1}{3}$.

The probability of choosing a penny is $\frac{1}{2}$ since there are now only 2 coins left. The probability of both is $\frac{1}{3} \cdot \frac{1}{2}$ or $\frac{1}{6}$.

Practice

Determine whether the events are independent or dependent.

1. selecting a marble and then choosing a second marble without replacing the first marble

2. rolling a number cube and spinning a spinner

3. Find the probability of each situation. A card is drawn from the cards at the right.

   a. $P(J \text{ or } K)$ **b.** $P(L \text{ or } M \text{ or } N)$ **c.** $P(L \text{ or a vowel})$

   [I | J | K | L]
   [M | N | O]

4. **Standardized Test Practice** David and Adrian have a coupon for a pizza with one topping. The choices of toppings are pepperoni, hamburger, sausage, onions, bell peppers, olives, and anchovies. If they choose at random, what is the probability that they both choose hamburger as a topping?

   A $\frac{1}{7}$ **B** $\frac{1}{49}$ **C** $\frac{2}{7}$ **D** $\frac{1}{42}$

Answers: 1. dependent; There is one less marble in the bag when the second marble is drawn. **2.** independent **3a.** $\frac{2}{7}$ **3b.** $\frac{3}{7}$ **3c.** $\frac{3}{7}$ **4.** B

© Glencoe/McGraw-Hill   106   Glencoe Pre-Algebra

NAME _____ DATE _____ PERIOD _____

12 Chapter Review

Heirloom Math

Use information about your family to complete the following.

1. Start by making an organized list of the names and ages of at least ten people in your immediate or extended family.

2. Make a stem-and-leaf plot of your data. Find the range, median, upper and lower quartiles, and the interquartile range for your data.

3. Now make a box-and-whisker plot of your data.

4. Refer to your data in Exercises 1–3. Which of these representations do you think best models your data and why?

5. Suppose your family is drawing names to exchange gifts. Each of the names in your data set are put into a hat.

    a. What is the probability of drawing the name of a person who is between 10 and 20 years old?

    b. What are the odds of drawing the name of a person who is older than 40?

    c. What is the probability that the first name drawn is yours?

    d. How many ways can the first three names be drawn?

Answers are located in the Answer Key.

© Glencoe/McGraw-Hill   107   Glencoe Pre-Algebra

NAME _____ DATE _____ PERIOD _____

13-1 Polynomials *(Pages 669–672)*

Expressions such as x^2 and $4ab$ are **monomials**. Monomials are numbers, variables, or products of numbers and variables. An algebraic expression that contains one or more monomials is called a **polynomial**. A polynomial is a sum or difference of monomials. A polynomial with two terms is called a **binomial**, and a polynomial with three terms is called a **trinomial**. The *degree* of a monomial is the sum of the exponents of its variables. A monomial like 3 that does not have a variable associated with it is called a **constant**. The degree of a nonzero constant is 0. The constant zero has no degree. The degree of a polynomial is the same as that of the term with the greatest degree.

Examples

Monomial or Polynomial	Variables	Exponents	Degree
y	y	1	1
$4z^3$	z	3	3
$5a^2b^3$	a, b	2, 3	2 + 3 or 5
12	none	none	0
$7q^2 + 2q + 1$	q	2, 1	2

Remember that $y = y^1$.

Practice

Classify each polynomial as a *monomial*, *binomial*, or *trinomial*.

1. $7x$
2. $k + 2$
3. $c^4 + 7$
4. $a^2 + a + 10$
5. $4xyz$
6. $m + 15$
7. $5 + 3a^2 + a$
8. $n + 18 + n^5$
9. $(-11)^2 - x + x^2$

Find the degree of each polynomial.

10. $9a^2 + 6$
11. $5x + 3$
12. 113
13. $p + p^3 + p^2$
14. $x^7 + x^5 + x^2$
15. $b^5 + 2b + 5b^3$

Evaluate each polynomial if $x = 5$, $y = -1$, and $z = -3$.

16. $6z + 3 + x$
17. $xy^2 + z + 5$
18. $-5yz + 2z$

19. **Recreation** A school recreation yard is to be built on an empty lot near the science classrooms. The perimeter of the yard is to be a rectangle with a width of x feet and a length that is 50 ft greater than the width. Write a polynomial that expresses the perimeter of the recreation yard.

20. **Standardized Test Practice** Find the degree of the polynomial $3x^5 + 6x^2 - 8x^7 + x^3 - 6$.

A 2   B 3   C 5   D 7

Answers: 1. monomial 2. binomial 3. binomial 4. trinomial 5. monomial 6. binomial 7. trinomial 8. trinomial 9. trinomial 10. 2 11. 1 12. 0 13. 3 14. 7 15. 5 16. −10 17. 7 18. −21 19. $4x + 100$ 20. D

© Glencoe/McGraw-Hill   Glencoe Pre-Algebra

13-2 Adding Polynomials (Pages 674–677)

The numerical part of a monomial is called the **coefficient**. For example, the coefficient of $-3y^4$ is -3. A monomial without a number in front of it, such as x^2, has a coefficient of 1, or, in the case of $-xy^2$, -1. When monomials are the same or differ only by their coefficients, they are called **like terms**. For example, a, $2a$, and $10a$ are all like terms. To add polynomials, combine like terms.

Examples   Find each sum.

a. $(3y + 2) + (6y + 9)$
You can add vertically. Align the like terms, then add.

$3y + 2$
$+6y + 9$
$\overline{\quad 9y + 11}$

b. $(4z + 8) + (2z - 5)$
Add horizontally. Use the associative and commutative properties to group like terms.

$(4z + 8) + (2z - 5)$
$= (4z + 2z) + (8 - 5)$ Group like terms.
$= (4 + 2)z + (8 - 5)$ Distributive Property
$= 6z + 3$ Simplify.

Try These Together

Find each sum.

1. $(3x + 2a) + (x + 3a)$
2. $(2m + 4) + (6 + 6m)$
3. $(g + h) + (g - h)$

HINT: Group like terms, then add.

Practice

Find each sum.

4. $(2x + 9) + (5x - 7)$
5. $(10x + 2y) + 3x$
6. $(4x - 6) + (x + 3)$
7. $b + (2x - 2b)$
8. $(3k^2 + 2m) + (m + 8)$
9. $(5x^2 + 2y) + (6y^2 + 3)$
10. $(3z^2 + 4 + z) + (2z + 6 + 5z^2)$
11. $(7x^2 + 3x - 2) + (5x^2 - 2x + 5)$
12. $(2k^3 + k^2 + k) + (3k^3 + 2k^2 + 4k + 5)$
13. $(5x^5 + 3x^2 + x) + (2x^3 + 3x^4 + 1)$

Find each sum. Then, evaluate if $x = 2$ and $y = -3$.

14. $(x^2 + xy + 3) + (x^2 + xy + 2)$
15. $(2x + xy + 6) + (y - xy + 2)$

16. **Art** Marta wants to frame two paintings. One has a perimeter of $5w + 3$ and the other has a perimeter of $7w + 4$. Write an expression for the total length of framing material Marta will need to frame these two paintings.

17. **Standardized Test Practice** Find $(3x^2 + 4y^2 + 2x) + (x^2 - 2y^2 + 7)$. Then, evaluate if $x = 4$ and $y = 5$.
    A 64
    B 114
    C 129
    D 132

Answers: 1. $4x + 5a$ 2. $8m + 10$ 3. $2g$ 4. $7x + 2$ 5. $13x + 2y$ 6. $5x - 3$ 7. $2x - b$ 8. $3k^2 + 3m + 8$ 9. $5x^2 + 6y^2 + 2y + 3$ 10. $8z^2 + 3z + 10$ 11. $12x^2 + x + 3$ 12. $5k^3 + 3k^2 + 5k + 5$ 13. $5x^5 + 3x^4 + 2x^3 + 3x^2 + x + 1$ 14. $2x^2 + 2xy + 5$; 1 15. $2x + y + 8$; 9 16. $12w + 7$ 17. C

13-3 Subtracting Polynomials (Pages 678–681)

Recall that you can subtract a rational number by adding its additive inverse. You can also subtract a polynomial by adding its additive inverse. To find the additive inverse of a polynomial, multiply the entire polynomial by −1, which effectively changes the sign of each term in the polynomial.

Examples Find each difference.

a. $(9y + 7) - (4y + 6)$

To subtract vertically, align the like terms and then subtract.

$$\begin{array}{r} 9y + 7 \\ (-)\ 4y + 6 \\ \hline 5y + 1 \end{array}$$

b. $(6z + 2) - (5z - 8)$

To subtract horizontally, add the additive inverse of the second polynomial.

$(6z + 2) - (5z - 8)$
$= (6z + 2) + (-1)(5z - 8)$
$= 6z + 2 + (-5z + 8)$ Group like terms.
$= (6z - 5z) + (2 + 8)$
$= 1z + 10$ or $z + 10$

Try These Together
Find each difference.

1. $(3t + 2) - (2t + 1)$
2. $(-2y + 4) - (10y + 3)$
3. $(6x + 7) - (8x + 4)$

Practice

State the additive inverse of each polynomial.

4. $8xy$
5. $k^2 + 7k$
6. $-3m + n - 7n^2$

Find each difference.

7. $(-9g - 2) - (-3g + 5)$
8. $(-11x + 4) - (3x + 2)$
9. $(6x - 3y) - (2x - 2y)$
10. $(5a - 12b) - (3a - 13b)$
11. $(4x^2 - 3) - (2x^2 + 5)$
12. $(c^2 + 7) - (c^2 - 5)$
13. $(6r^2 + 8r - 3) - (2r^2 + 4r - 1)$
14. $(5b^2 + 3b - 15) - (-3b^2 + 4b - 2)$
15. $(7m^2 - 4m - 5) - (-2m^2 - 3m - 3)$
16. $(7x^3 - 2x^2 + 4x + 9) - (5x^3 - 2x^2 - x + 4)$

17. **Geometry** The perimeter of the trapezoid is $8x + 18$. Find the missing length of the lower base.

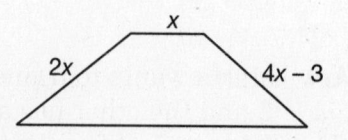

18. **Standardized Test Practice** Find the difference of $10x^3 + 4x^2 - 6x + 15$ and $5x^3 - 2x^2 - 5x - 3$.

   A $5x^3 + 6x^2 - x + 18$
   B $-5x^3 - 6x^2 - x + 18$
   C $15x^3 + 2x^2 + x + 18$
   D $-15x^3 - 6x^2 - x + 18$

Answers: 1. $t + 1$ 2. $-12y + 1$ 3. $-2x + 3$ 4. $-8xy$ 5. $-k^2 - 7k$ 6. $3m - n + 7n^2$ 7. $-6g - 7$ 8. $-14x + 2$ 9. $4x - y$ 10. $2a + b$ 11. $2x^2 - 8$ 12. 12 13. $4r^2 + 4r - 2$ 14. $8b^2 - b - 13$ 15. $9m^2 - m - 2$ 16. $2x^3 + 5x + 5$ 17. $x + 21$ 18. A

13-4 Multiplying a Polynomial by a Monomial

(Pages 683–686)

You can use the distributive property to multiply a polynomial by a monomial.

Examples Find each product.

a. $5(x^2 + 2x + 1)$

$5(x^2 + 2x + 1) = 5(x^2) + 5(2x) + 5(1)$ Distributive Property
$= 5x^2 + 10x + 5$ Multiply.

b. $3d(2d - 8)$

$3d(2d - 8) = 3d(2d) - 3d(8)$ Distributive Property
$= 6d^2 - 24d$ Multiply monomials.

Try These Together

Find each product.

1. $6(2x + 3)$
2. $4(z + 4)$
3. $2x(x^2 + 3x - 5)$

HINT: Use the Distributive Property to multiply every term in the polynomial by the monomial.

Practice

Find each product.

4. $2z(z - 4)$
5. $-5v(1 + v)$
6. $m(m - 6)$
7. $5b(-12 + 2b)$
8. $-2x(3x - 7x)$
9. $x(y^2 + z)$
10. $-2x(4 - 4y + 6y^2)$
11. $3b(b^3 + b^2 + 5)$
12. $-5x(2x^3 + 2x^2 - 4)$
13. $3d(d^4 + 5d^3 + 6)$
14. $s(s^2 - 2s^3 + 7)$
15. $7(-8x + 5x^2 + y^2)$

Solve each equation.

16. $6(2z + 10) + 8 = 5z + 5$
17. $-3(x - 4) = 4x + 8$
18. $2(6y - 11) = 5y + 3$
19. $5(-2x + 8) = -6x + 20$

20. **Woodshop** Devonte is making a wooden box for a project in woodshop. The base of the box has width x inches and length $x + 5$ inches. What polynomial represents the area of the base of the box?

21. **Standardized Test Practice** Find the product of a and $a + b + c^2$.

A $a + ab + ac$ B $a^2 + ab + ac$ C $a^2 + ab + ac^2$ D $a + b^2 + ac$

Answers: 1. $12x + 18$ 2. $4z + 16$ 3. $2x^3 + 6x^2 - 10x$ 4. $2z^2 - 8z$ 5. $-5v - 5v^2$ 6. $m^2 - 6m$ 7. $-60b + 10b^2$ 8. $8x^2$ 9. $xy^2 + xz$ 10. $-8x + 8xy - 12xy^2$ 11. $3b^4 + 3b^3 + 15b$ 12. $-10x^4 - 10x^3 + 20x$ 13. $3d^5 + 15d^4 + 18d$ 14. $s^3 - 2s^4 + 7s$ 15. $-56x + 35x^2 + 7y^2$ 16. -9 17. $\frac{4}{7}$ 18. $3\frac{4}{7}$ 19. 5 20. $x^2 + 5x$ 21. C

© Glencoe/McGraw-Hill   111   Glencoe Pre-Algebra

NAME _____ DATE _____ PERIOD ___

13-5 Linear and Nonlinear Functions

(Pages 687–691)

As you may recall, an equation whose graph is a straight line is called a linear function. A linear function has an equation that can be written in the form of $y = mx + b$. Equations whose graphs are not straight lines are called **nonlinear functions**. Some nonlinear functions have specific names. A **quadratic function** is nonlinear and has an equation in the form of $y = ax^2 + bx + c$, where $a \neq 0$. Another nonlinear function is a **cubic function**. A cubic function has an equation in the form of $y = ax^3 + bx^2 + cx + d$, where $a \neq 0$.

Function	Equation	Graph
Linear	$y = mx + b$	
Quadratic	$y = ax^2 + bx + c, a \neq 0$	
Cubic	$y = ax^3 + bx^2 + cx + d, a \neq 0$	

Examples Determine whether the function is linear or nonlinear.

a. $y = 4x$

Linear, $y = 4x$ can be written as $y = mx + b$.

b. $y = x^2 + x - 2$

Nonlinear, $y = x^2 + x - 2$ cannot be written as $y = mx + b$

c. $y = \dfrac{7}{x}$

Nonlinear, $y = \dfrac{7}{x}$ cannot be written as $y = mx + b$.

Practice

Determine whether the function is linear or nonlinear.

1. $y = 5$
2. $2x + 3y = 10$
3. $y = 7x^2$
4. $xy = -13$

5. **Standardized Test Practice** Select the nonlinear function.

A $y = -3x - 5$ B $y = 0.75$ C $y = 3x + x^2$ D $y = \dfrac{1}{2}x + 2$

Answers: 1. linear 2. linear 3. nonlinear 4. nonlinear 5. C

© Glencoe/McGraw-Hill   112   Glencoe Pre-Algebra

NAME _____ DATE _____ PERIOD _____

13-6 Graphing Quadratic and Cubic Functions

(Pages 692–696)

You can graph quadratic functions and cubic functions using a table of values.

Examples Make a table of values, plot the points, and connect the points using a curve to graph each equation.

a. $y = 0.5x^2$

x	$y = 0.5x^2$	(x, y)
−2	$y = 0.5 \cdot (-2)^2$	(−2, 2)
−1	$y = 0.5 \cdot (-1)^2$	(−1, 0.5)
0	$y = 0.5 \cdot (0)^2$	(0, 0)
1	$y = 0.5 \cdot (1)^2$	(1, 0.5)
2	$y = 0.5 \cdot (2)^2$	(2, 2)

b. $y = x^3 + x$

x	$y = x^3 + x$	(x, y)
−2	$y = (-2)^3 + (-2)$	(−2, −10)
−1	$y = (-1)^3 + (-1)$	(−1, −2)
0	$y = (0)^3 + 0$	(0, 0)
1	$y = (1)^3 + 1$	(1, 2)
2	$y = (2)^3 + 2$	(2, 10)

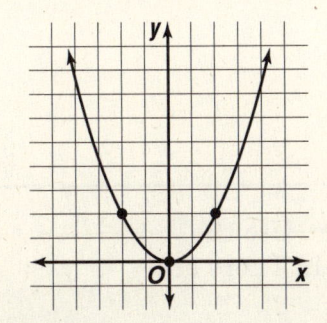

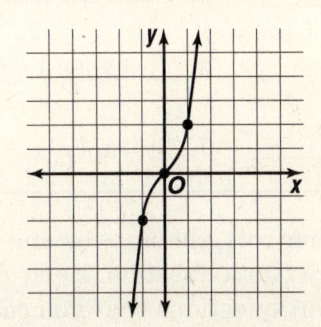

Practice

Graph each equation.

1. $y = x^2$
2. $y = x^3$
3. $y = x^2 - 2$
4. $y = x^3 - 1$
5. $y = x^3 + 2x$
6. $y = -2x^2$
7. $y = x^2 + 3$
8. $y = -2x^3 + 1$

9. **Standardized Test Practice** Which equation is represented by the graph at the right.

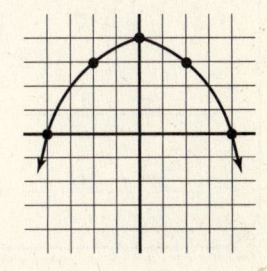

A $y = x^2 + 4$
B $y = -x^3 + 4$
C $y = -x^2 + 2.75$
D $y = -0.25x^2 + 4$

Answers: 1.–8. See Answer Key 113 9. D

© Glencoe/McGraw-Hill   113   Glencoe Pre-Algebra

NAME _____ DATE _____ PERIOD _____

13 Chapter Review

Role Playing

For this review you will play the roles of both student and teacher. In the student role, you will answer each question. In the teacher role, you will write each question.

Student Role

_____ 1. What is the degree of the polynomial $y^4 - y^8 + 100$?

_____ 2. Simplify $(2x^3)^5$.

_____ 3. Simplify $3x(x^4)^2$.

_____ 4. Add $(4x + 5y) + (y - 3x)$.

_____ 5. Subtract $(2a + 7b) - (8a - b + 1)$.

_____ 6. Multiply $3x(4x + 5)$.

_____ 7. Multiply $(x + 2)(x + 4)$.

For the teacher role, the answer and some keywords are given. Use the keywords to write a question. Refer to questions 1–7 as models. There are many different questions that you can write for each answer.

Teacher Role

___4___ 8. _____
(keywords: degree of a polynomial)

___$4x + 9$___ 9. _____
(keywords: adding polynomials)

___$-x + 7$___ 10. _____
(keywords: adding polynomials)

___$3x + 1$___ 11. _____
(keywords: subtracting polynomials)

___$x^2 - 16$___ 12. _____
(keywords: multiplying polynomials)

Answers are located in the Answer Key.

© Glencoe/McGraw-Hill   114   Glencoe Pre-Algebra

Answer Key

Lesson 1-1

1a. You know the number of species in each group. You need to find the total number of species.   **b.** Add the numbers for all groups.   **c.** The total is 4,888,288 species.   **d.** Round the number of species in each group to the nearest thousand and add. This gives an estimate of 4,889,000. This is close to the calculated answer. So the answer seems reasonable.

Chapter 1 Review

1. 9   2. −10   3. 36   4. −3x + 4
5. 5x − 2
Drawing: 🍌 + (🍎 + 🍓)

Lesson 2-1

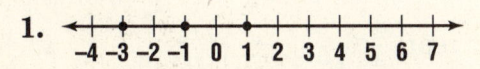

1. number line with points at −3, 0, 2, 5
2. number line with points at −4, −1, 2, 4
3. number line with points at −2, 1, 3, 6
4. number line with points at −3, 0, 3, 5

Chapter 2 Review

1st Play: 12; 28
2nd Play: −5; 33
3rd Play: 18; 15
4th Play: 16; −1
Yes. The negative number, −1, signifies a touchdown.

Chapter 3 Review

1. 40   2. −50   3. 1200   4. 3850
5. 1925   6. 1975

Mrs. Acevedo was born in 1975, so subtract that year from the current year to find her age.

Chapter 4 Review

ACROSS 1. $6ab^3$ 3. $\frac{2a}{b^2}$ 4. $\frac{1}{81}$ 5. $\frac{4}{7}$
8. $\frac{5x}{y^3}$ 10. 56   12. $\frac{x^4}{6y}$ 13. $48mn$
15. 30
DOWN 1. $60a^4$ 2. 22   3. $21x^3y^4$ 6. 7^5
7. 15   9. x^2y^3 11. $6mn$ 12. x^6 14. 8^3

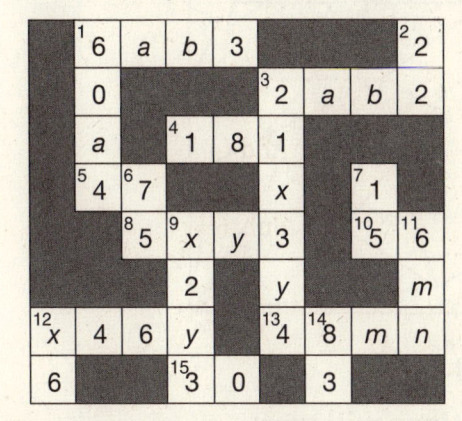

Chapter 5 Review

1. Andrew: $a = 0.3$; Nancy: $n = 0.25$; Jocelyn: $j = \frac{2}{5}$; Samantha: $s = \frac{1}{10}$; Mark: $m = \frac{1}{20}$

2. $\frac{1}{20}, \frac{1}{10}, \frac{1}{4}, \frac{3}{10}, \frac{2}{5}$

3. Jocelyn ate the most, and Mark ate the least.   4. Drawings may vary so long as sizes of each slice are correct relative to each other.

Chapter 6 Review

1–15. Sample answers are given.
1. Kelton   2. 3 out of 4   3. Steve
4. 2.5   5. Jack   6. $6.75   7. Monique
8. 2 out of 5   9. Kelton   10. Kelton
11. 0.3   12. 90%   13. 0.4   14. 75%
15. $\frac{9}{10}$ 16. $14.40   17. 1020 were male.
18a. 17.5%   18b. 82.5%

© Glencoe/McGraw-Hill   Glencoe Pre-Algebra

Answer Key

Chapter 7 Review
1. $x < -1$ 2. $x = -8$ 3. $x = 4$
4. $x < 16$ 5. $x > 16$ 6. $x = 6$
7. $x = -27$
The hidden picture looks like this:

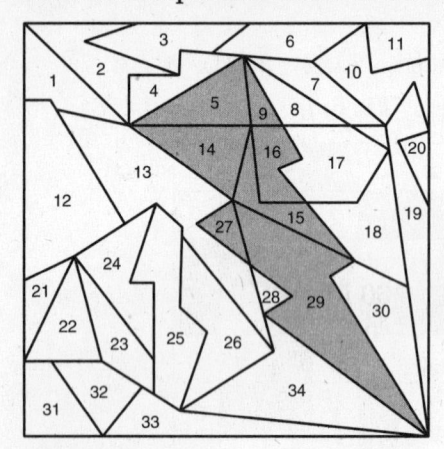

Lesson 8-2
4–6. Solutions will vary.
4. 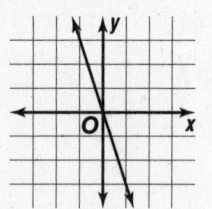
5. 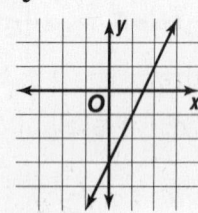
6.

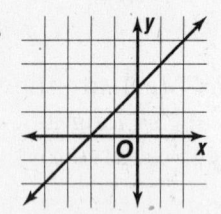

Lesson 8-3
1. x-intercept: $1\frac{1}{2}$;
   y-intercept: -3

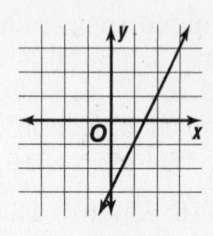

2. x-intercept: 1;
   y-intercept: 1

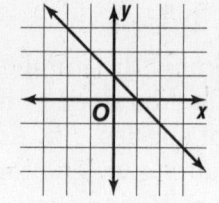

3. x-intercept: 6;
   y-intercept: -4

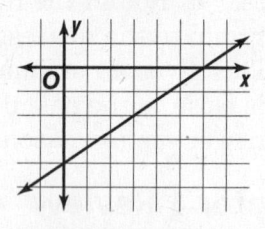

4. x-intercept: 4;
   y-intercept: 2

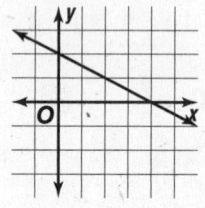

5. x-intercept: $\frac{2}{3}$;
   y-intercept: -2

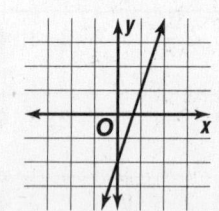

6. x-intercept: 2;
   y-intercept: 4

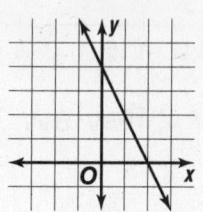

7. x-intercept: 3;
   y-intercept: 3

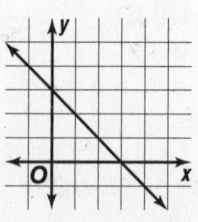

8. x-intercept: -6;
   y-intercept: 2

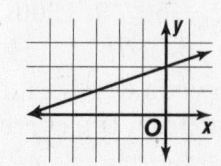

© Glencoe/McGraw-Hill

Glencoe Pre-Algebra

Answer Key

9. x-intercept: $\frac{1}{2}$;
   y-intercept: -1

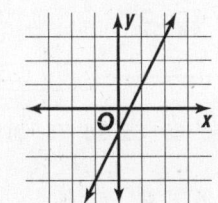

Lesson 8-9

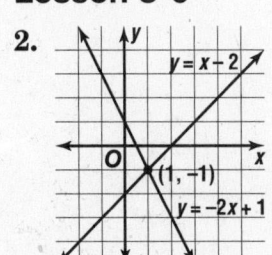

2. (graph showing $y = x - 2$ and $y = -2x + 1$ intersecting at $(1, -1)$)

Lesson 8-10

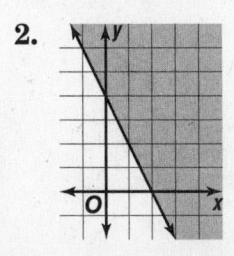

2. (graph)
6. (graph)
7. (graph)
8. (graph)
9. (graph)
10. (graph)
11. (graph)

Chapter 8 Review

1. $f(x)$ and $g(x)$ 2. $x = -3$ 3. $y = 0$
4. 1   5. $f(x)$ only   6. -4

The solution to the puzzle is BOILED EGGS.

Chapter 9 Review

1–5. Sample answers are given.
1. Equation: $80^2 + 30^2 = c^2$
   Solution: $c = 85.44$ in.
   Actual: 85.5 in.
2. Equation: $48^2 + 36^2 = c^2$
   Solution: $c = 60$ in.
   Actual: 36.13 in.
3. Equation: $16^2 + b^2 = 19^2$
   Solution: $b = 10.25$ in.
   Actual: 12 in.
4. Equation: $74^2 + b^2 = 80^2$
   Solution: $b = 30.40$ in.
   Actual: actual diagonal was 38 in.
5. The solutions were different from the actual measurements in most cases because it was hard to get an exact measurement, especially on the TV and bed.

Lesson 10-3

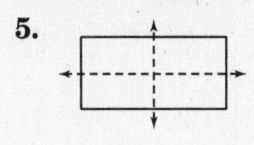

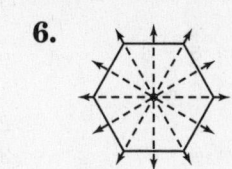

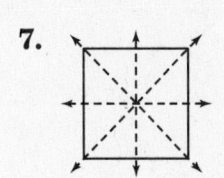

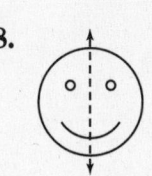

5. (rectangle with symmetry lines)
6. (hexagon with symmetry lines)
7. (square with symmetry lines)
8. (smiley face with symmetry line)

© Glencoe/McGraw-Hill   117   Glencoe Pre-Algebra

Answer Key

Chapter 10 Review

136	38	45	115
59	101	94	80
87	73	66	108
52	122	129	31

Sum = 334

Chapter 11 Review

Hat: 96 in^3
Head: 216 in^3
Neck: 9 in^3
Arm: 106 in^3
Torso: 1500 in^3
Leg: 226 in^3
Foot: 90 in^3

Total volume = 2665 in^3

Lesson 12-1

1.
1	2 2 8
2	2 7
3	3
4	2 3

    4|2 = 42

2.
8	1
9	1
10	5 9
11	4 5 9
12	0

    12|0 = 120

3.
5	1 3 7 9
6	1 3 8
7	
8	1 9
9	0

    9|0 = 9.0

4.
0	3 5 7
1	0 1 5 8
2	1 2
3	0

    3|0 = 30

5.
28	4
29	2
30	5 7 9
31	6
32	
33	2

    28|4 = $28,400

Median price: $30,700; Choice of the better representation will vary.

Lesson 12-3

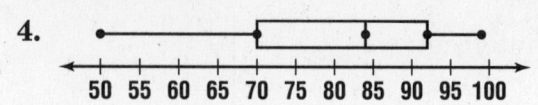

4. [box-and-whisker plot from 50 to 100, with points at 50, 70, 85, 92, 100]

Lesson 12-6

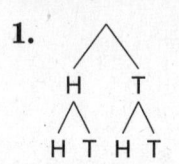

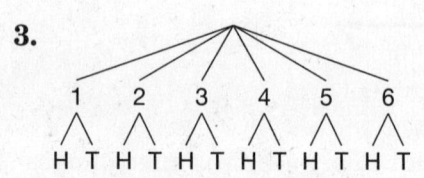

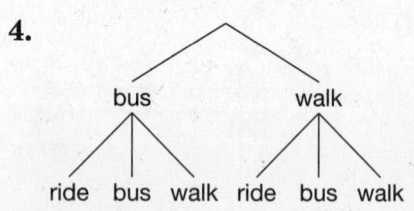

1. [tree diagram: H, T branching to H T H T]

3. [tree diagram: 1, 2, 3, 4, 5, 6 each branching to H T]

4. [tree diagram: bus, walk each branching to ride, bus, walk]

Chapter 12 Review

1–5. Sample answers are given.

1.

Name	Age
Mom	38
Dad	41
Me	13
Larry	8
Juanita	4
Grandma	63
Grandpa	68
Uncle Juan	25
Aunt Mary	30
Cousin Margarita	2

© Glencoe/McGraw-Hill

Glencoe Pre-Algebra

Answer Key

2.
```
6 | 3 8
5 |
4 | 1
3 | 0 8
2 | 5
1 | 3
0 | 2 4 8    2/5 = 25.
```

range: 66; median: 27.5; upper quartile: 41; lower quartile: 8; interquartile range: 33

3. [box-and-whisker plot from 0 to 70]

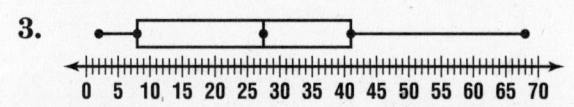

4. I think that the stem-and-leaf plot best models the data because it organizes the data so you can easily see the range of ages from least to greatest.

5a. $\frac{1}{10}$ **5b.** $\frac{3}{10}$ **5c.** $\frac{1}{10}$ **5d.** 9

Lesson 13-6

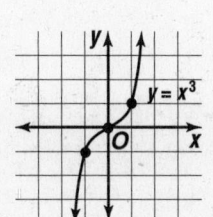

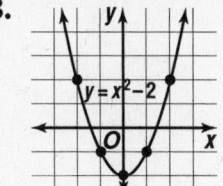

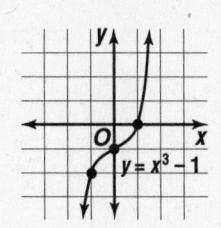

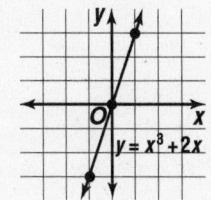

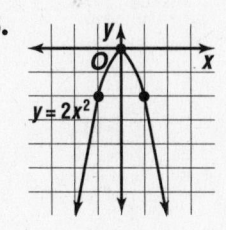

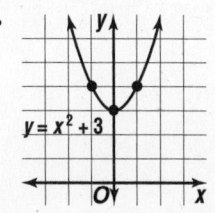

1. [graph of $y = x^2$] **2.** [graph of $y = x^3$]

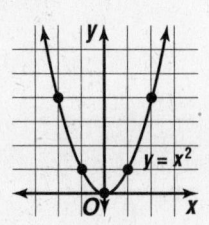

3. [graph of $y = x^2 - 2$] **4.** [graph of $y = x^3 - 1$]

5. [graph of $y = x^3 + 2x$] **6.** [graph of $y = 2x^2$]

7. [graph of $y = x^2 + 3$] **8.** [graph of $y = 2x^3 + 1$]

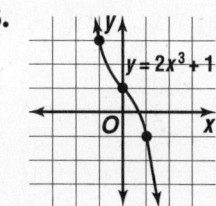

Chapter 13 Review

1. 8   **2.** $32x^{15}$ **3.** $3x^9$ **4.** $x + 6y$
5. $-6a + 8b - 1$ **6.** $12x^2 + 15x$
7. $x^2 + 6x + 8$ **8.** The student needs to supply a polynomial with a degree of 4. To find the degree of a polynomial, you must find the degree of each term. The greatest degree of any term is the degree of the polynomial. Sample answer: $x^2 + 2y^4$ has a degree of 4 because the first term has a degree of 2 and the second term has a degree of 4; since 4 is greater, the degree of the polynomial is 4.   **9.** The student needs to supply two polynomials that when added, have a sum of $4x + 9$. To add polynomials, you add the like terms. Sample answer: $(3x + 5) + (x + 4)$; In this sentence, $3x + x = 4x$ and $5 + 4 = 9$.   **10.** The student needs to supply two polynomials that when added, have a sum of $-x + 7$. To add polynomials, you add the like terms. Sample answer: $(2x + 6) + (-3x + 1)$
11. The student needs to supply two polynomials that when subtracted, have a difference of $3x + 1$. To subtract polynomials, you subtract the like terms. Sample answer: $(6x + 5) - (3x + 4)$
12. The student needs to supply two polynomials that when multiplied, have a product of $x^2 - 16$. Sample answer: $(x + 4)(x - 4)$

© Glencoe/McGraw-Hill   Glencoe Pre-Algebra

The temperate forests of Australia.

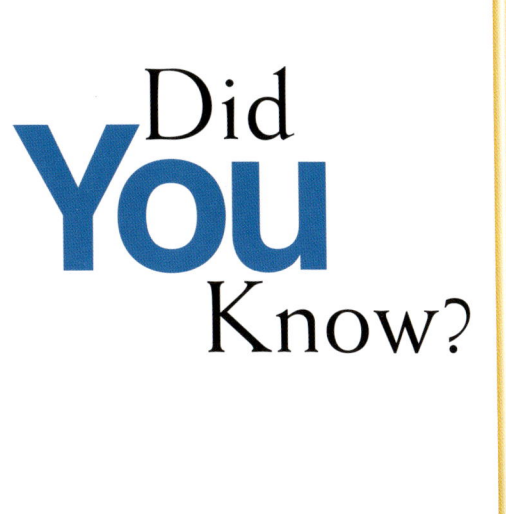

Did You Know?

- Australia is one of the largest countries in the world.

- Australia is the smallest and the flattest continent in the world.

- Only 14% of Australia is forest; most of the land is desert!

- Australia's temperate forests are found along its coasts.

- Australia, like other countries, has many laws to limit commercial hunting of birds and other animals.

For Fr. James D. Ciupek (Ciup), a Kingston-like comforter
Viirginia Kroll

For my kids, Wolf and Teal
Michael S. Maydak

Message to Parents

Bear & Company, part of The Boyds Collection, Ltd. family, is committed to creating quality reading and play experiences that inspire kids to learn, imagine, and explore the world around them. Working together with experienced children's authors, illustrators, and educators, we promise to create stories and products that are respectful to your children and that will earn your respect in turn.

Published by Bear & Company Publications
Copyright © 2002 by Bear & Company

All rights reserved. No part of this book may be reproduced or used in any manner whatsoever without written permission. For information regarding permission, write to:
 Permissions
 Bear & Company Publications
 P.O. Box 3876
 Gettysburg, PA 17325

Printed in the United States of America
My Home™ is a registered trademark of Bear & Company.

Based on a series concept by Dawn Jones
Edited by Dawn Jones
Designed by Vernon Thornblad

Library of Congress Cataloging-in-Publication Data
Kroll, Virginia L.
 Kingston's flowering forest / written by Virginia Kroll ; illustrated by Michael S. Maydak.
  p. cm. -- (My home ; 3)
"Based on a series concept by Dawn L. Jones."
Summary: A gray koala envies a lorikeet's more colorful appearance and plans to make himself rainbow-colored, too, until he learns the value of blending in.
 ISBN 0-9712840-5-9 (alk. paper)
 [1. Animal defenses--Fiction. 2. Koala--Fiction. 3. Animals--Australia--Fiction. 4. Protective coloration (Biology)--Fiction. 5. Color--Fiction. 6. Australia--Fiction.] I. Maydak, Michael S., ill. II. Jones, Dawn L., 1963- III. Title.
PZ7.K9227 Ki 2001
[E]--dc21
                          2001005151

Kingston's Flowering Forest

By Virginia Kroll   Illustrated by Michael S. Maydak

bear & company

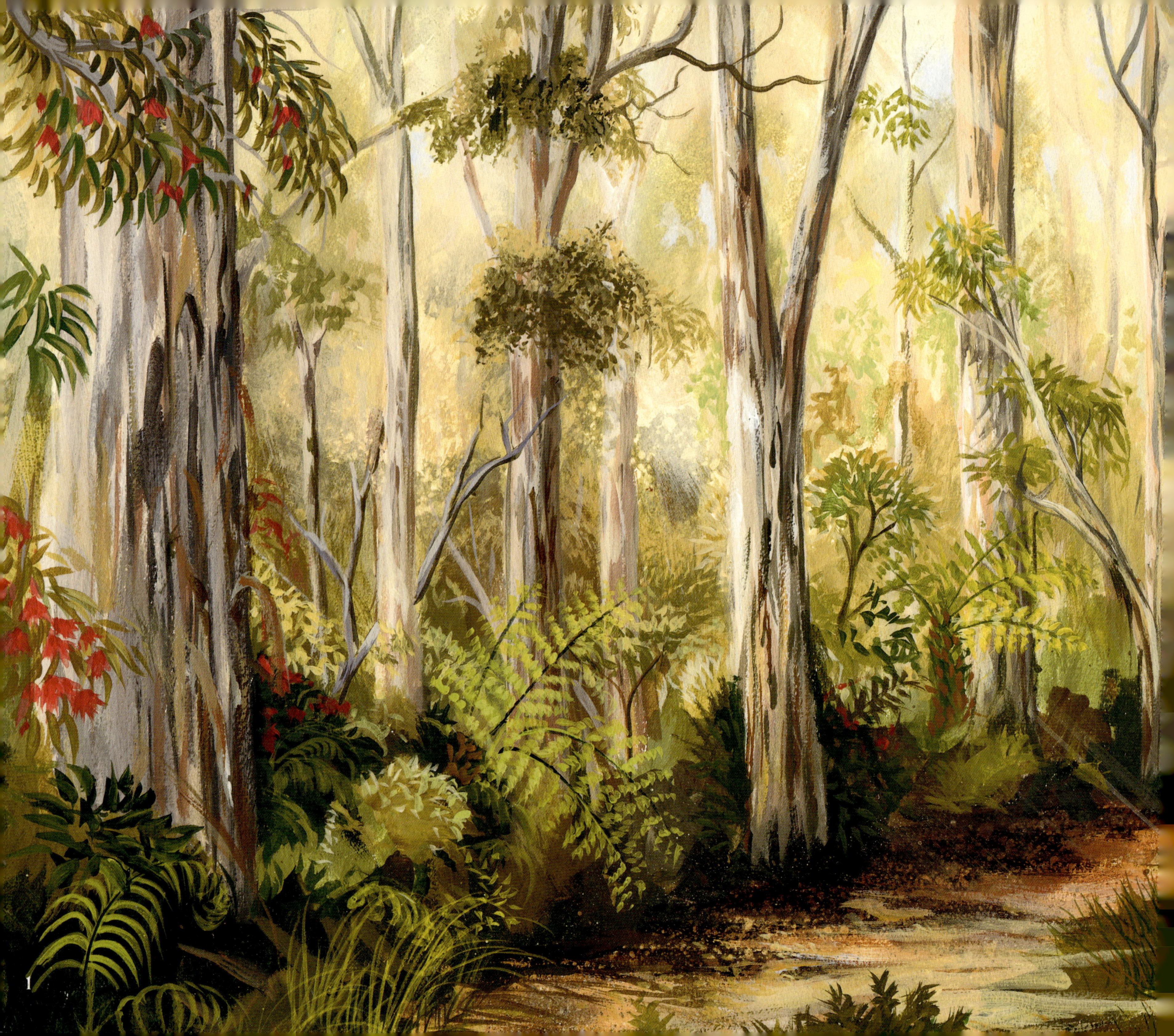

1

On the island of Australia, south of the equator, there is a place where colors burst like fireworks in the sun—red bottlebrushes and waratah flowers…golden wattle…orchids of every color. Some animals hide in the tall trees, eating leaves, fruits, and colorful flowers. But hiding isn't easy for everyone….

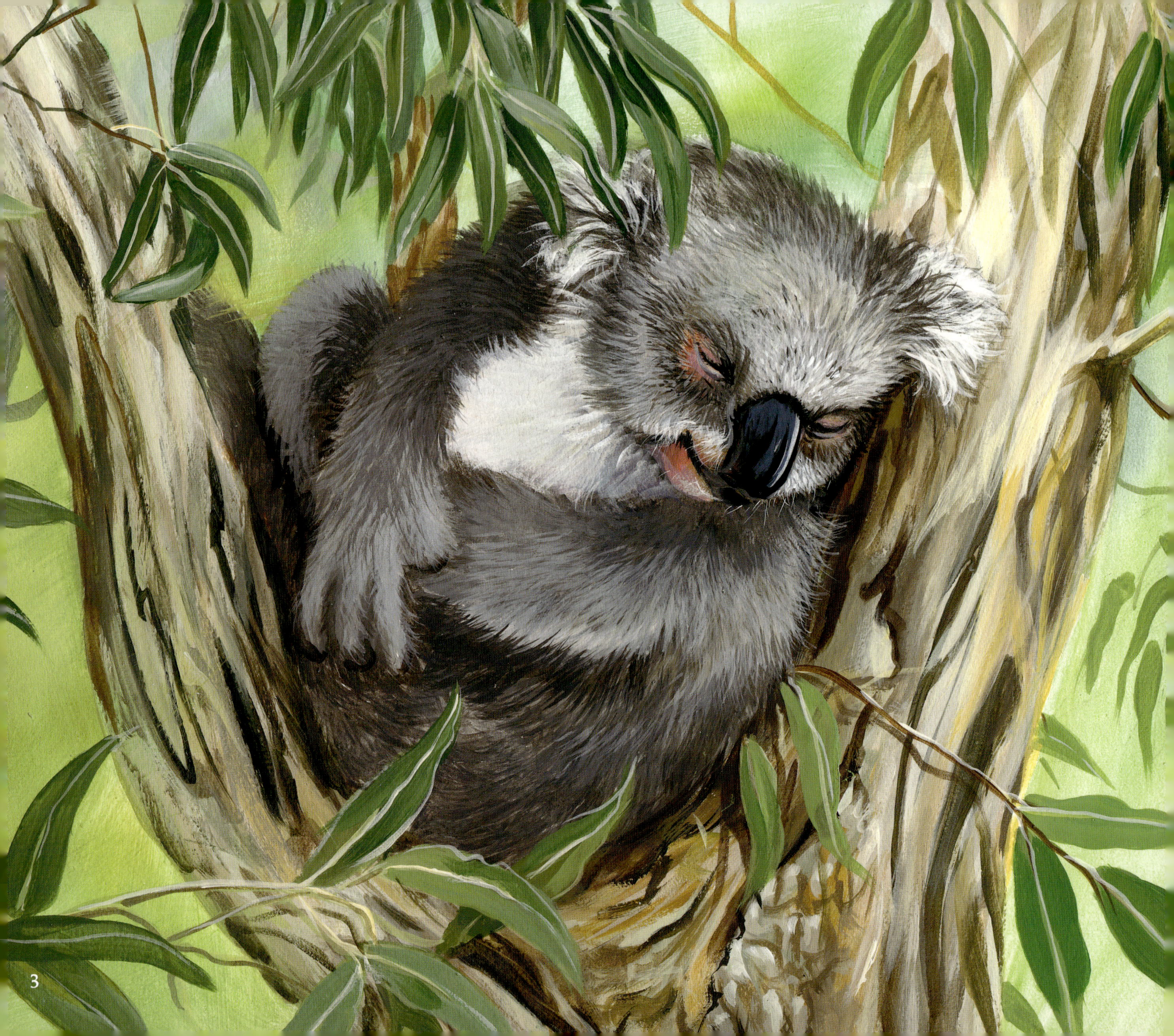

One dew-drenched Eastern Australia morning, Kingston the koala was curled up in a ball of gray, fast asleep. He slept until a terrible screeching made him uncoil his fluffy self.

Kingston yawned, stretched, and opened his small, oval eyes. He was usually good at ignoring noise. But the bird cries that echoed from saltbush to swamp gum tree were very loud. "What's all that racket?" he mumbled.

Sweet Tooth the flying fox was roosting with her family nearby. She answered, "Cooper the kookaburra and Rainbow the lorikeet are fighting again. I hope they stop soon, because my family and I want to sleep until dusk, when our favorite flowers open."

Kingston knew who Cooper the kookaburra was. He had heard the bird's call. It sounded like a wild laugh. But he didn't know Rainbow. Kingston wanted to ask Sweet Tooth about the lorikeet, but the flying fox had fallen back to sleep.

Kingston decided to find out for himself. He flexed his strong-clawed paws. Then he climbed, bottom first, down the gray gum tree. He could hear noisy chatter coming from a nearby mallee shrub.

Kingston waddled on all fours toward the noise. Suddenly—OUCH! The koala tripped! He rolled right onto something that looked like a prickly, poky, black-and-tan pincushion.

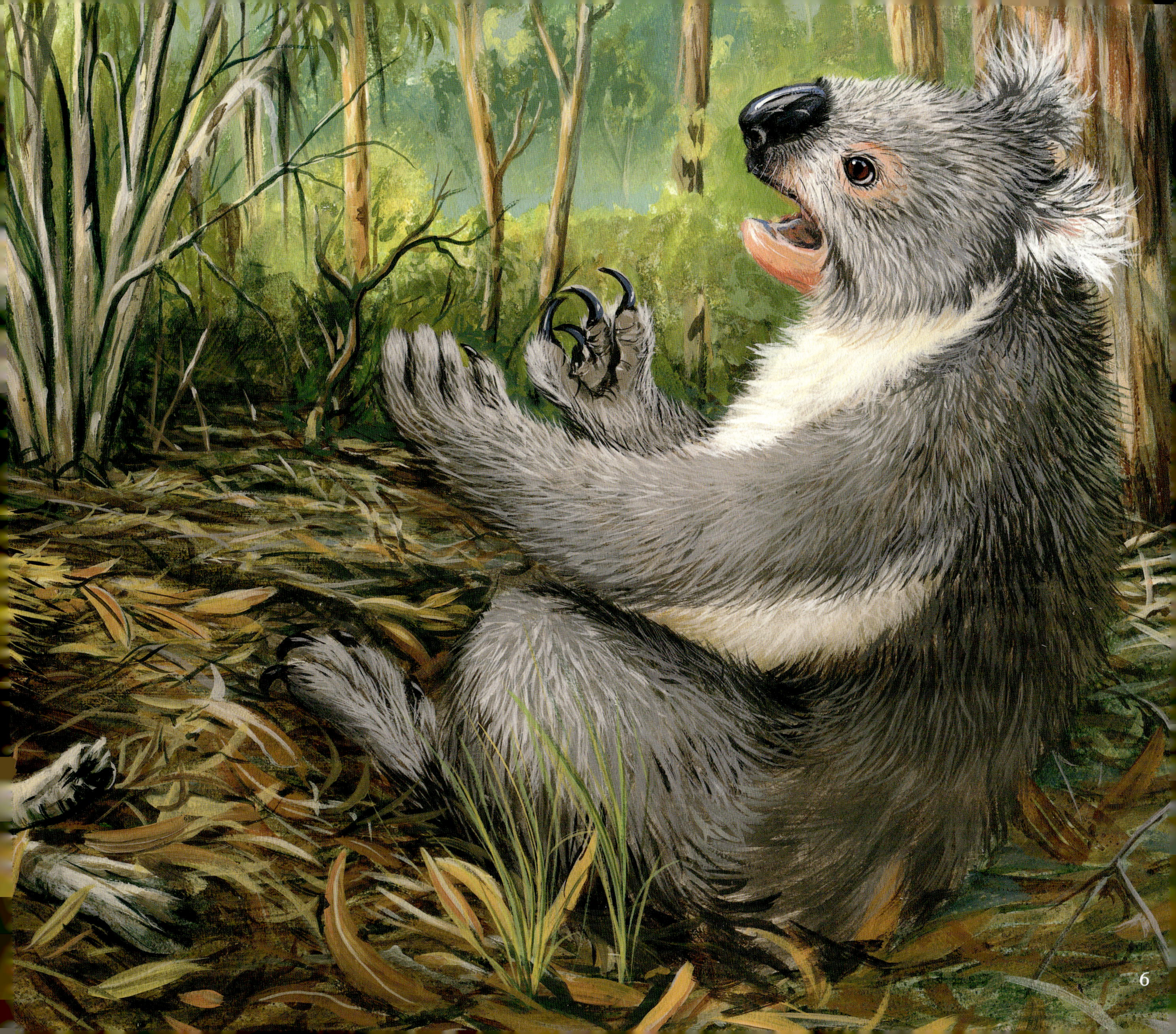

"Hello," it said, lifting its long snout and blinking its black eyes. "I'm Adelaide the echidna."

Kingston stared. He had never seen an animal like Adelaide.

"Did I hurt you?" Adelaide asked.

"A little," Kingston admitted.

"I'm sorry," said Adelaide. "It can be dangerous down here on the ground. So I protect myself by digging down and poking out my spines. When those birds started squawking, I was afraid a hunter was near. So I dug down and poked out."

8

9

A big-tailed, large-footed wallaby leaped toward Kingston and Adelaide. "You're lucky you can dig and poke," she said. "I have to jump to safety."

"Hi, Victoria," said Adelaide. "I think you're lucky to be tall and speedy."

"I'm slower now that my pouch is full," Victoria said. Baby Byron peeked out.

She pulled up a clump of tender grass—roots and all—for Baby Byron. Then she grabbed a twig for herself. "Yummy bark," she said. "Would you like some?"

Adelaide sniffed and shook her head. "I'm off to find some appetizing ants and tasty termites," she said. She eyed the tallest of the nearby termite towers.

"Not for me. I only eat gum-tree leaves," Kingston said.

The birds started squawking again. Byron disappeared into his mother's pouch as Victoria and Kingston went toward the noise.

As soon as Kingston saw Rainbow, he froze. She was the most beautiful creature he had ever seen. At first, all he could do was stare. But then he remembered why he had come.

"What are you two fighting about?"

"That!" said the birds. They both nodded toward a bright-red tree orchid.

"Cooper, you don't even eat flowers," said Victoria.

"But *I* do," said Rainbow. "And I want a bite of that blossom!"

"I want the beetles that are on it!" cried Cooper.

Kingston scratched his head thoughtfully. "I have an idea," he said. "Why doesn't Cooper carefully nibble the bugs off the blossom and leave the petals for Rainbow?"

Cooper and Rainbow cocked their feathered heads and looked at each other.

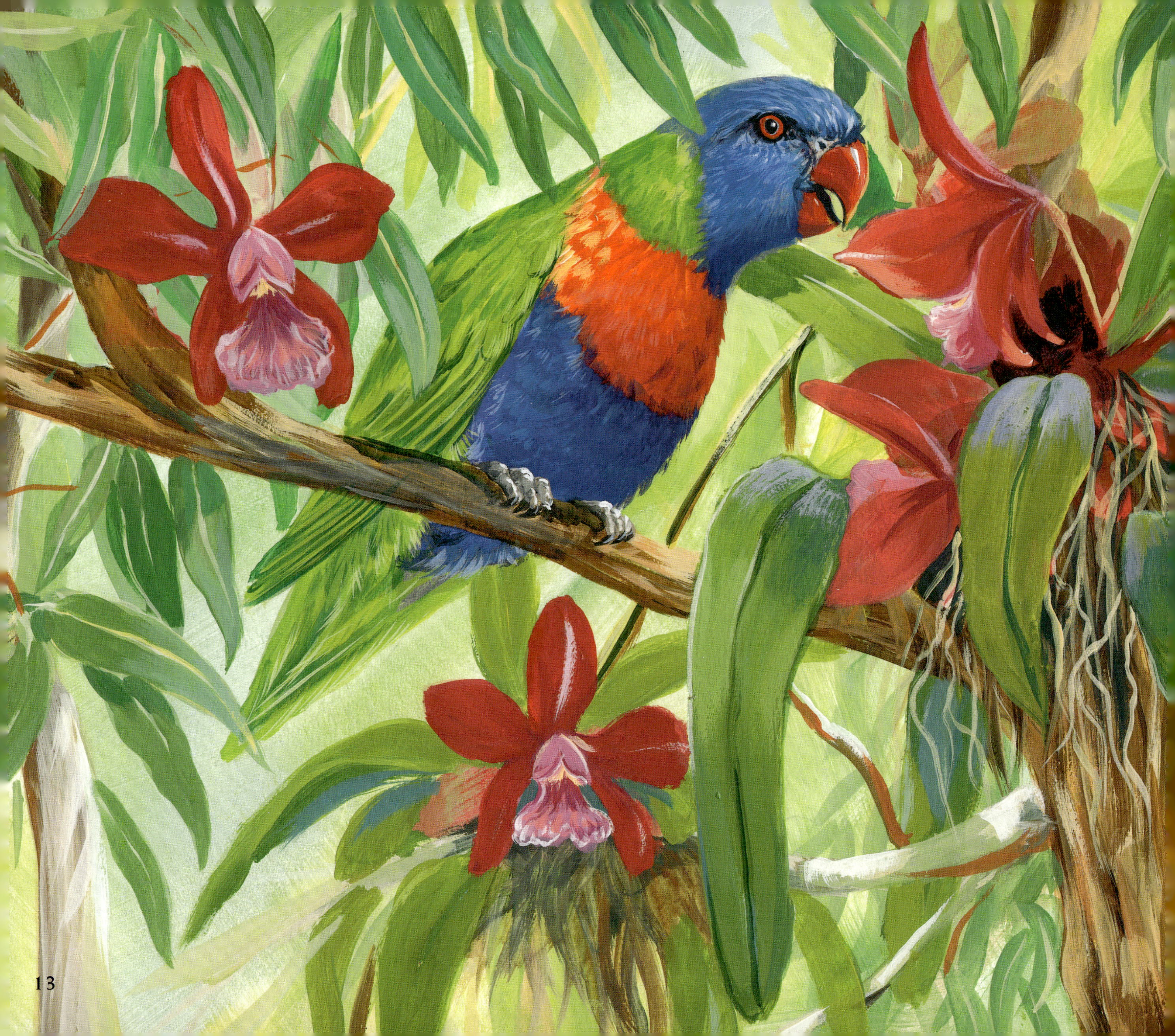

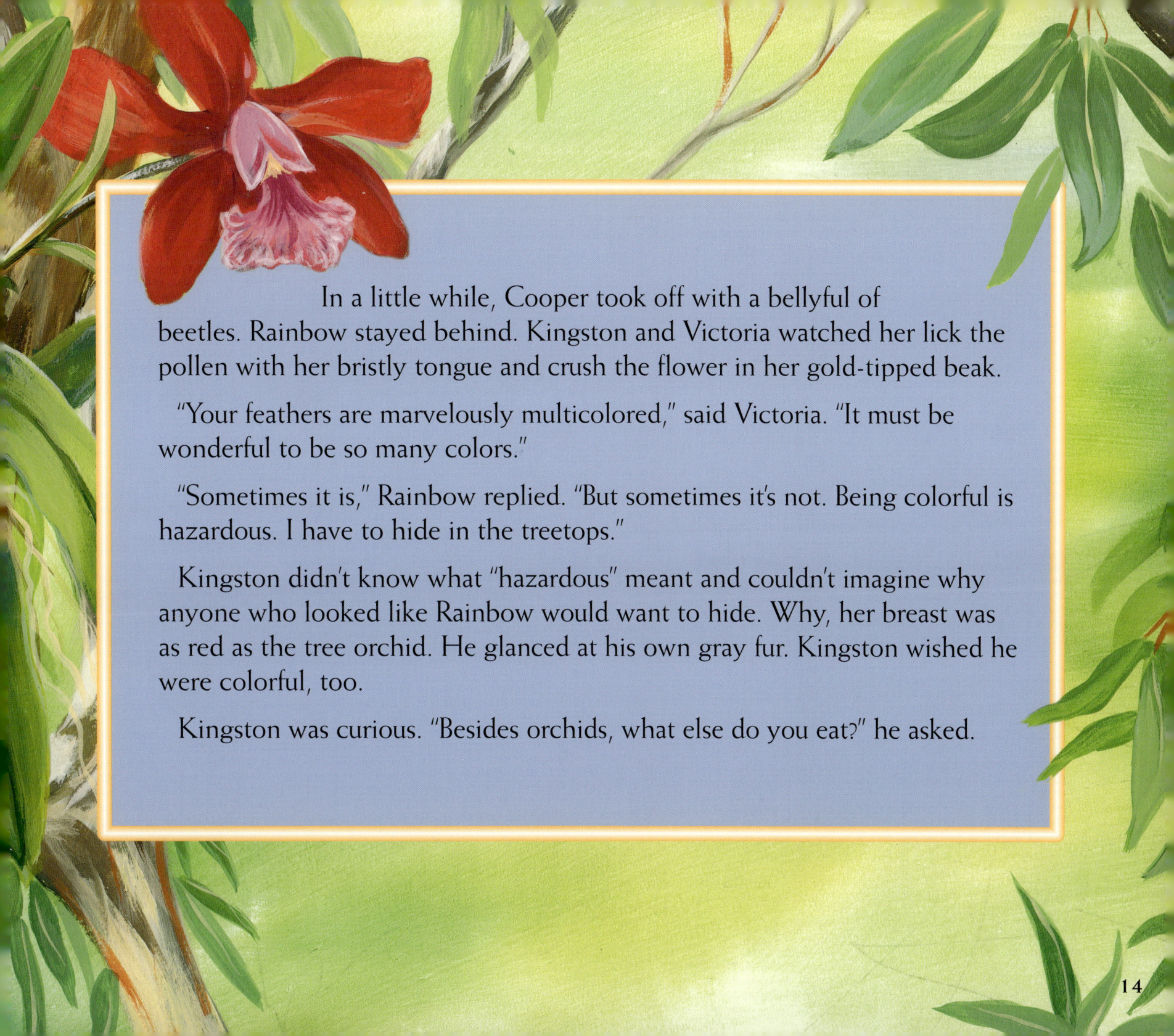

In a little while, Cooper took off with a bellyful of beetles. Rainbow stayed behind. Kingston and Victoria watched her lick the pollen with her bristly tongue and crush the flower in her gold-tipped beak.

"Your feathers are marvelously multicolored," said Victoria. "It must be wonderful to be so many colors."

"Sometimes it is," Rainbow replied. "But sometimes it's not. Being colorful is hazardous. I have to hide in the treetops."

Kingston didn't know what "hazardous" meant and couldn't imagine why anyone who looked like Rainbow would want to hide. Why, her breast was as red as the tree orchid. He glanced at his own gray fur. Kingston wished he were colorful, too.

Kingston was curious. "Besides orchids, what else do you eat?" he asked.

"Red waratahs, yellow bananas, and purple plums, to name a few more. I eat any flower or fruit that catches my eye." Then she flew away.

"That's it!" Kingston cried. "I'm going to change my diet!"

Victoria frowned. "What are you talking about? You just told me you only eat gum-tree leaves."

"I know," said Kingston. "But, don't you see? That's why I'm gray! All my life, all I've eaten are gray gum-tree leaves. I've never tasted the leaves of the Sydney blue gum or the river red gum or the Wallengary white gum. I'll bet if I did, I'd be marvelously multicolored, too."

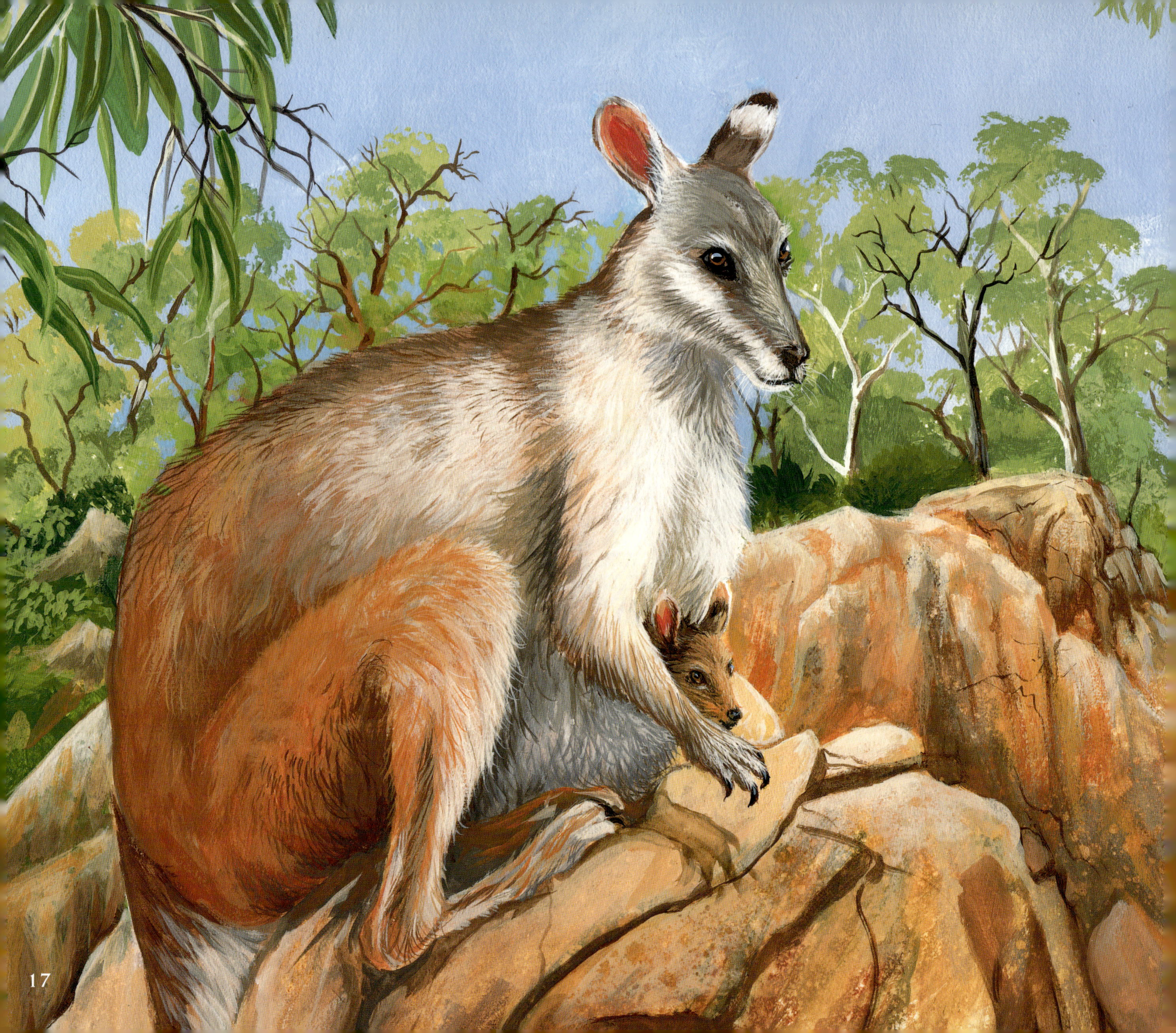

"What makes you think so?" Victoria asked. "Baby Byron and I eat gobs of green grass. But we're brown from head to toe."

"But you also eat tons of twigs and bundles of bark," said Kingston, "and they're all brown. Besides, think about Adelaide. She eats ants of black and termites of tan. And just look at her two colors."

"Well," said Victoria slowly, "maybe you do have a point."

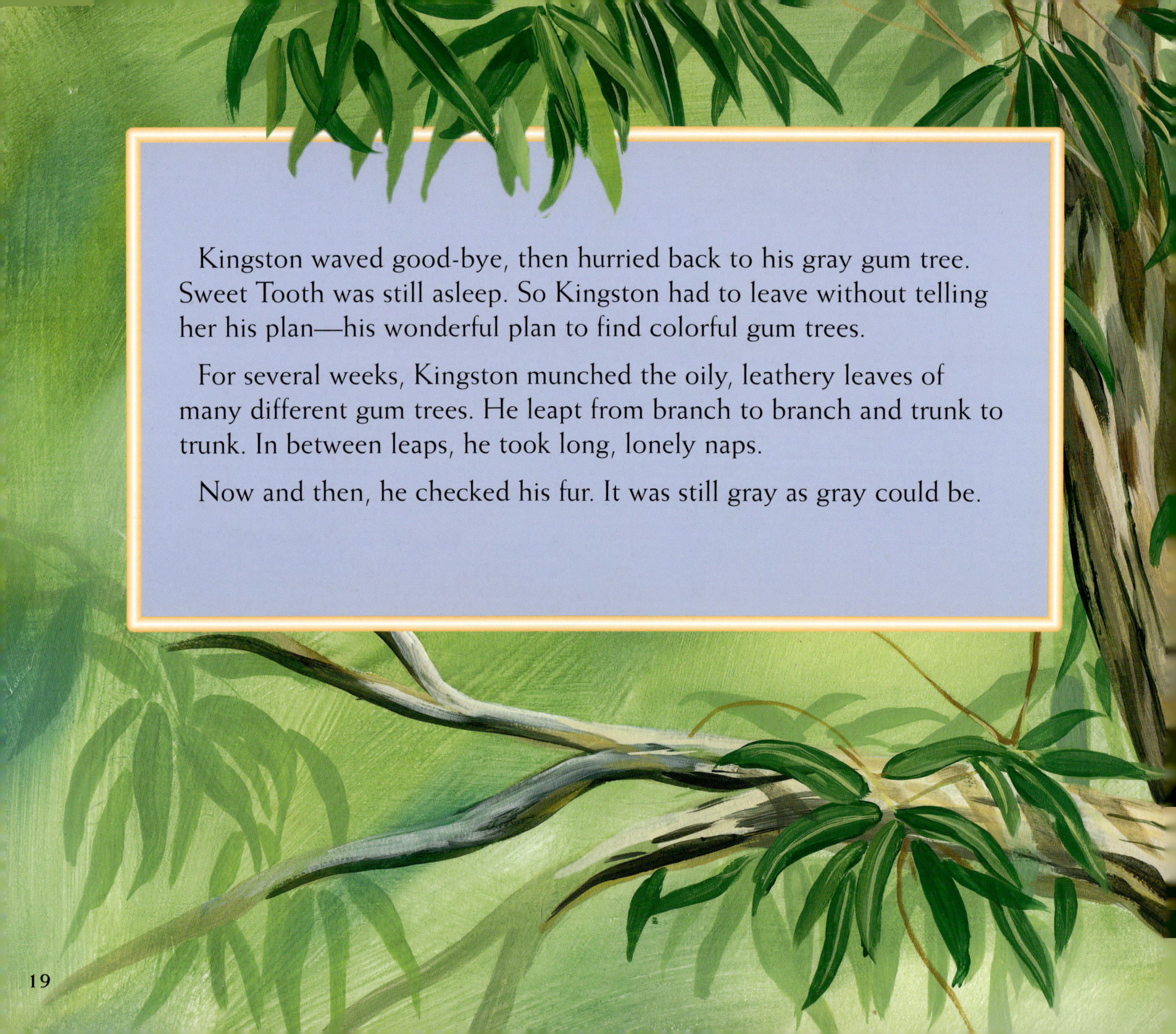

Kingston waved good-bye, then hurried back to his gray gum tree. Sweet Tooth was still asleep. So Kingston had to leave without telling her his plan—his wonderful plan to find colorful gum trees.

For several weeks, Kingston munched the oily, leathery leaves of many different gum trees. He leapt from branch to branch and trunk to trunk. In between leaps, he took long, lonely naps.

Now and then, he checked his fur. It was still gray as gray could be.

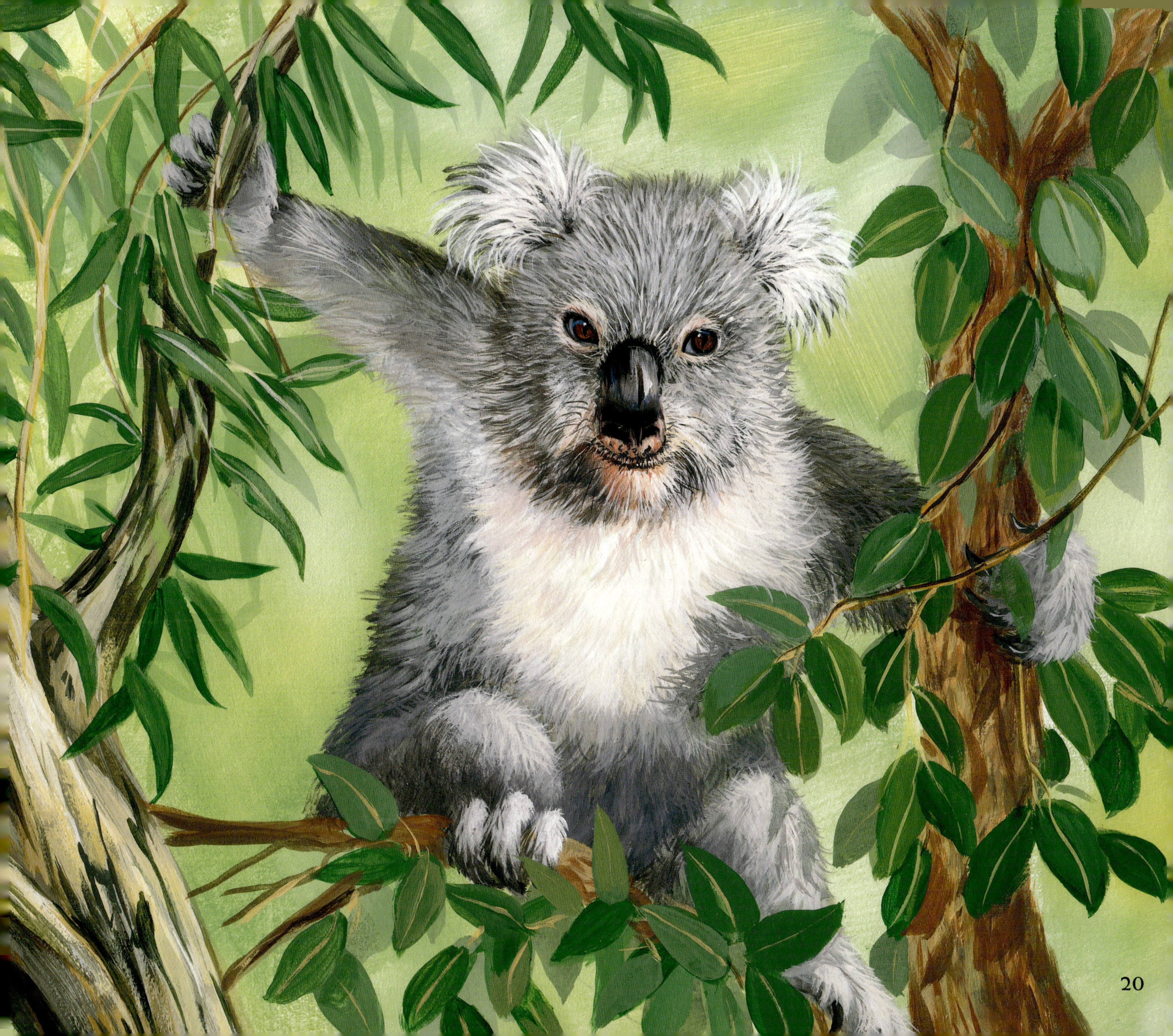

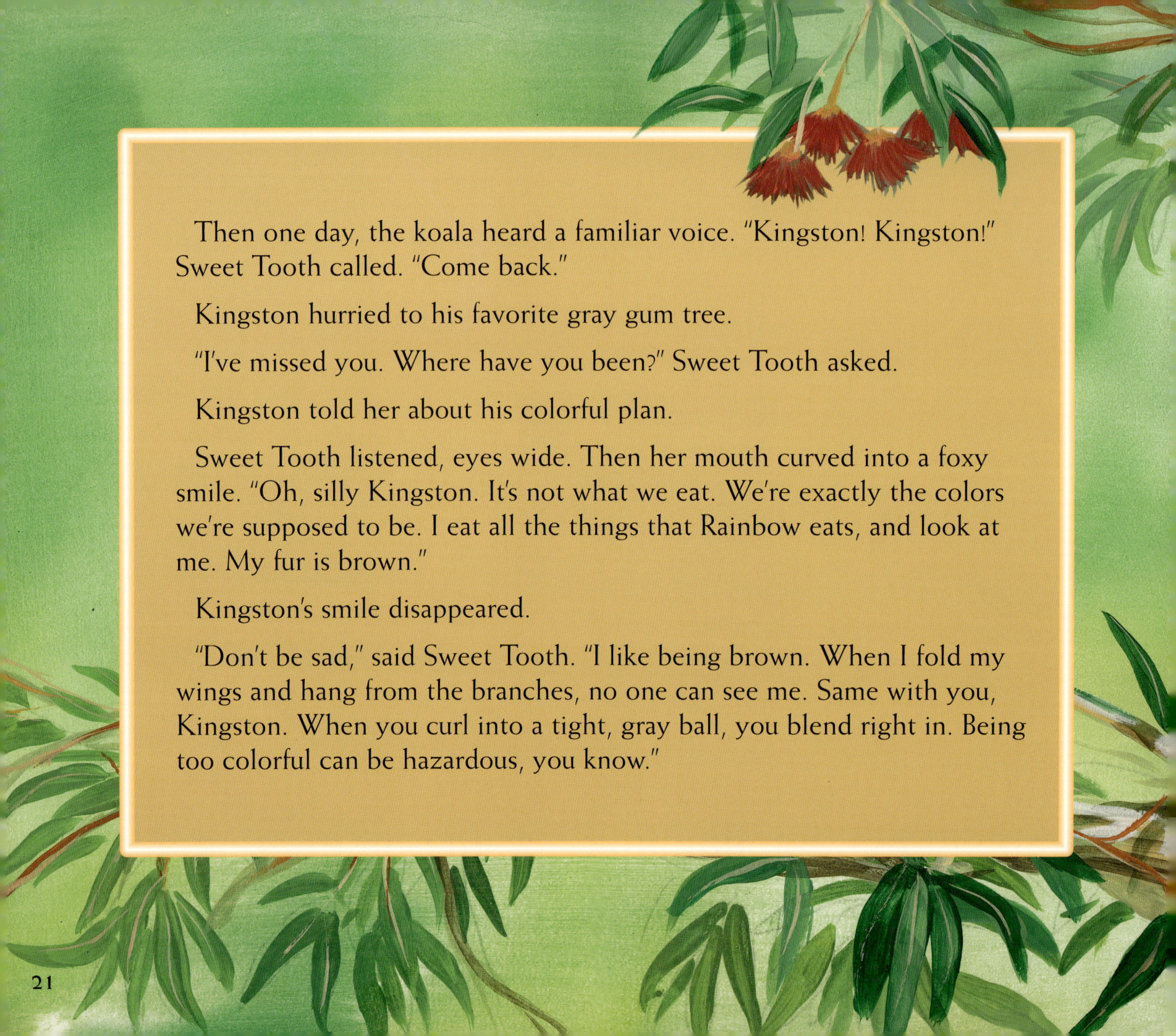

Then one day, the koala heard a familiar voice. "Kingston! Kingston!" Sweet Tooth called. "Come back."

Kingston hurried to his favorite gray gum tree.

"I've missed you. Where have you been?" Sweet Tooth asked.

Kingston told her about his colorful plan.

Sweet Tooth listened, eyes wide. Then her mouth curved into a foxy smile. "Oh, silly Kingston. It's not what we eat. We're exactly the colors we're supposed to be. I eat all the things that Rainbow eats, and look at me. My fur is brown."

Kingston's smile disappeared.

"Don't be sad," said Sweet Tooth. "I like being brown. When I fold my wings and hang from the branches, no one can see me. Same with you, Kingston. When you curl into a tight, gray ball, you blend right in. Being too colorful can be hazardous, you know."

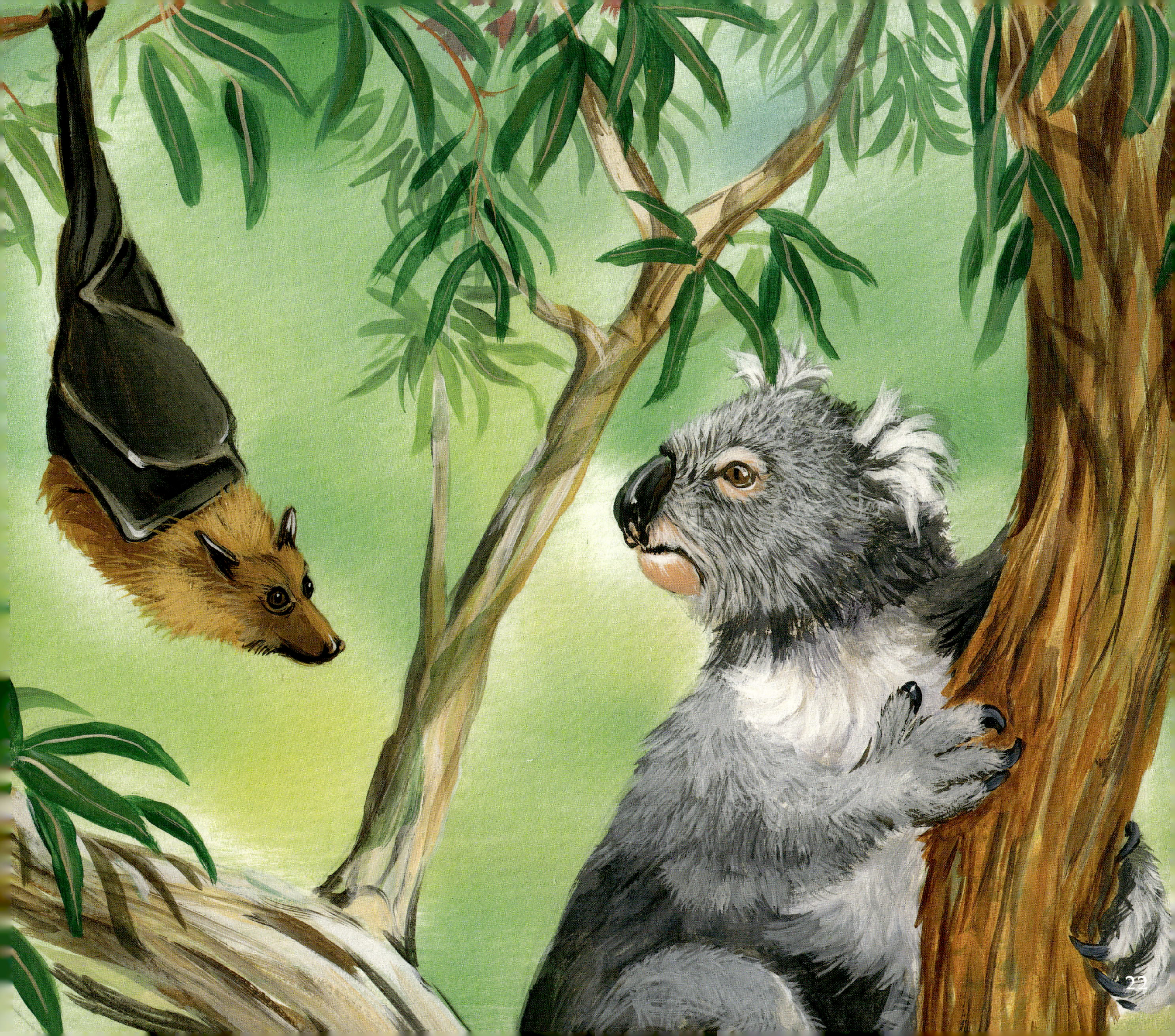

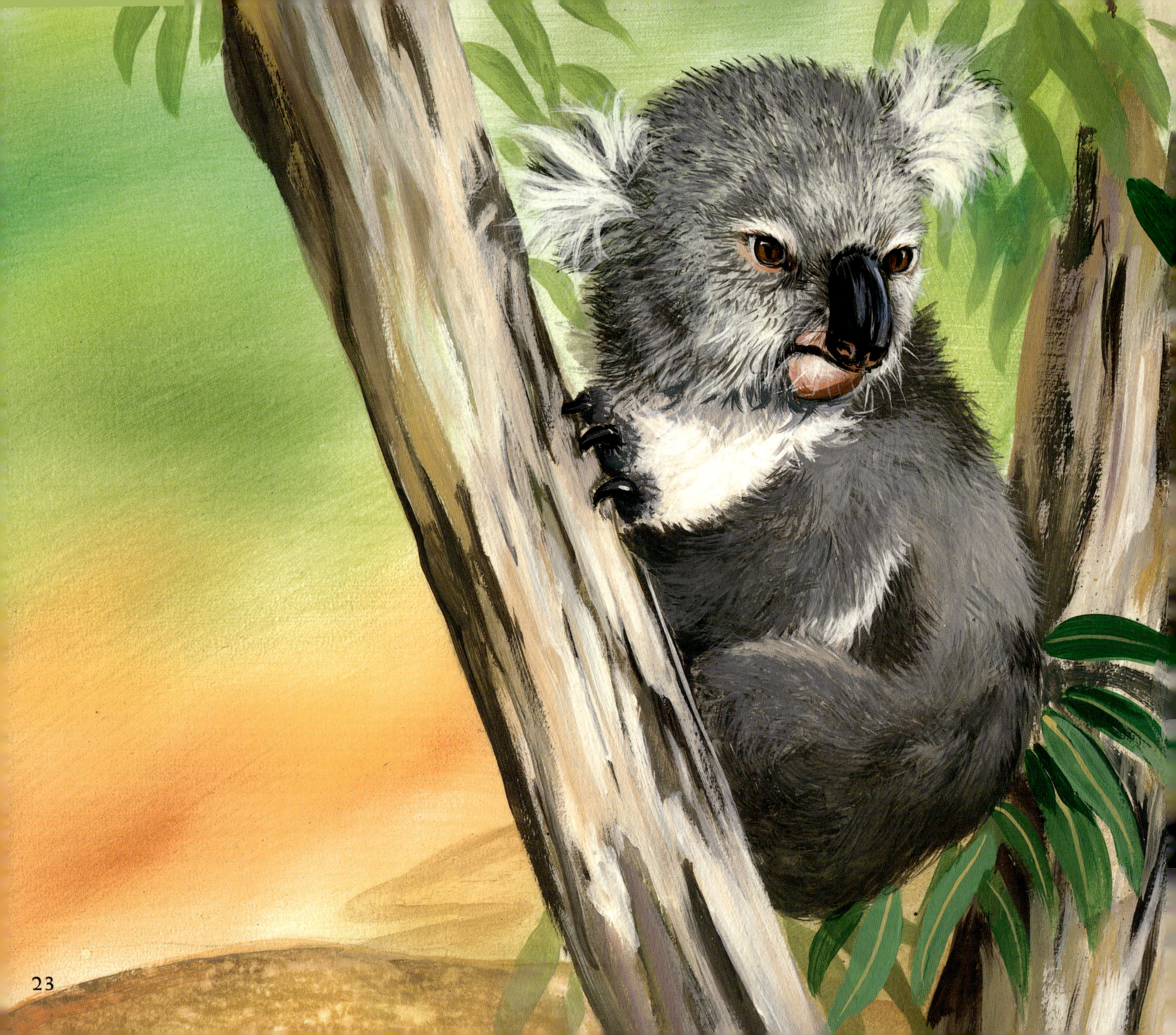

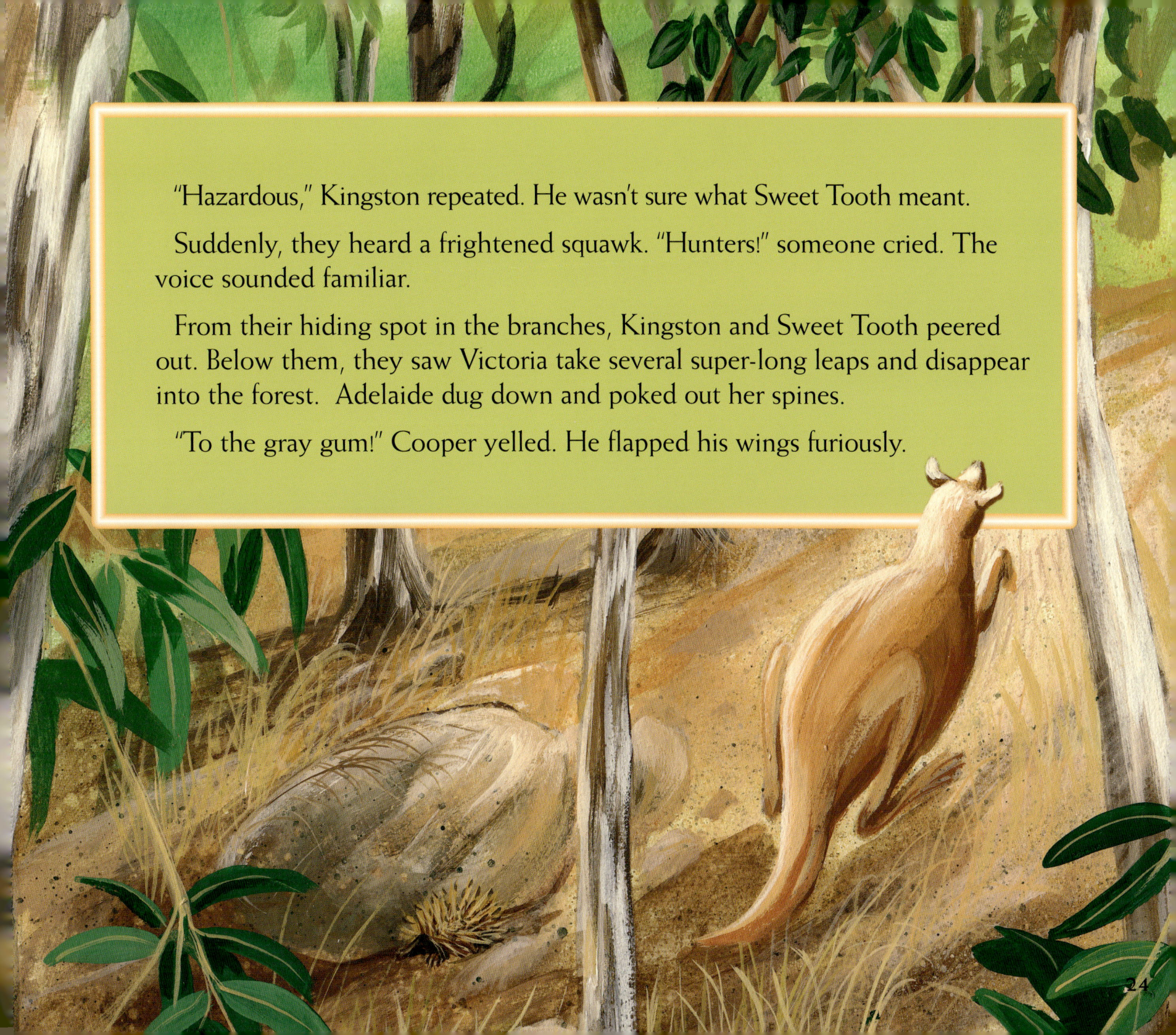

"Hazardous," Kingston repeated. He wasn't sure what Sweet Tooth meant.

Suddenly, they heard a frightened squawk. "Hunters!" someone cried. The voice sounded familiar.

From their hiding spot in the branches, Kingston and Sweet Tooth peered out. Below them, they saw Victoria take several super-long leaps and disappear into the forest. Adelaide dug down and poked out her spines.

"To the gray gum!" Cooper yelled. He flapped his wings furiously.

25

Suddenly, a bright streak of red, orange, yellow, green, blue, and purple broke through the branches. "Rainbow!" Kingston cried.

"Hunters!" cried Rainbow. "They're trying to trap me!"

Kingston knew what to do. "Come here!" he said as he patted his gray tummy. The bird flew to Kingston. She flattened her colorful feathers against his soft, gray fur.

27

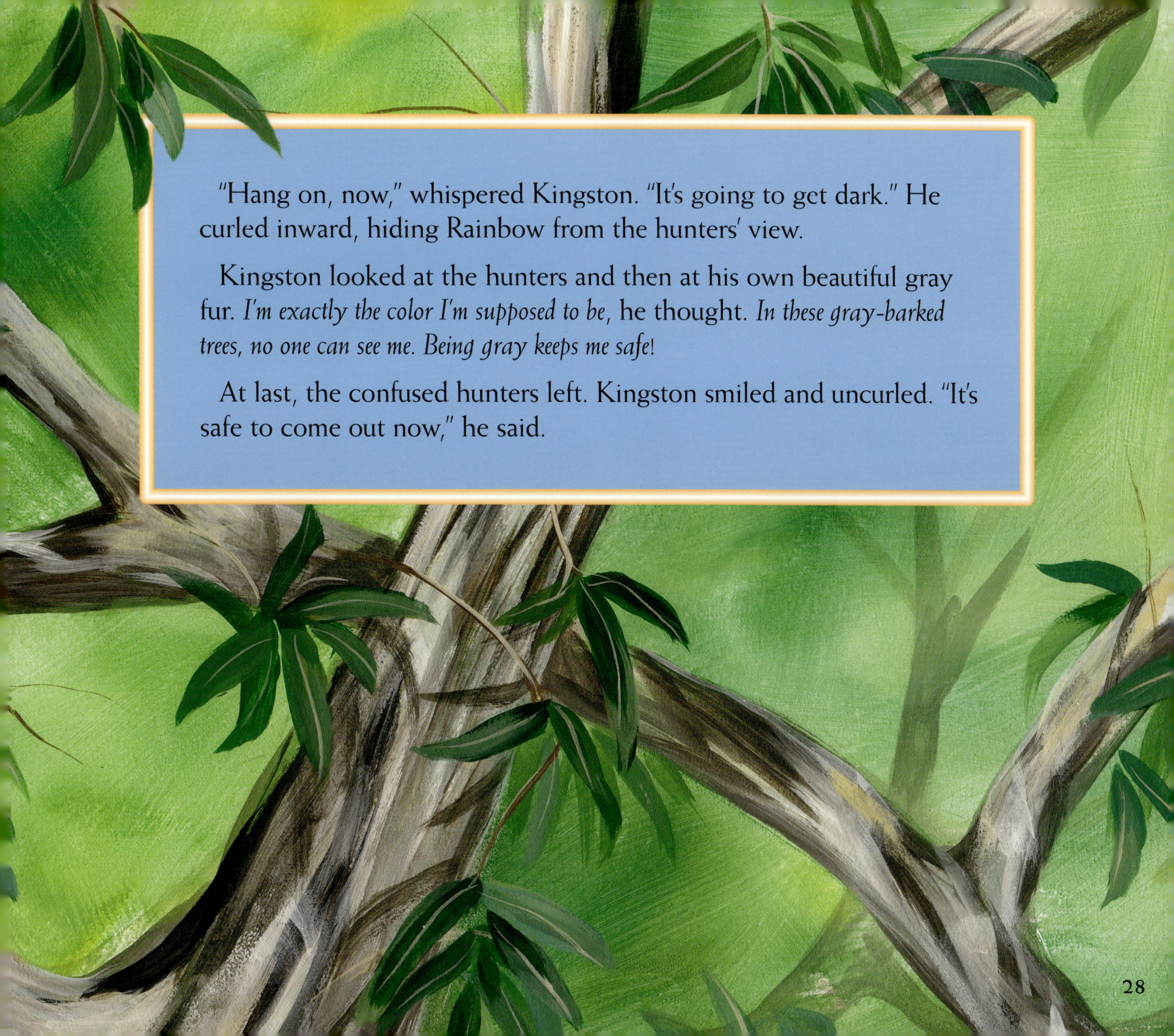

"Hang on, now," whispered Kingston. "It's going to get dark." He curled inward, hiding Rainbow from the hunters' view.

Kingston looked at the hunters and then at his own beautiful gray fur. *I'm exactly the color I'm supposed to be*, he thought. *In these gray-barked trees, no one can see me. Being gray keeps me safe!*

At last, the confused hunters left. Kingston smiled and uncurled. "It's safe to come out now," he said.

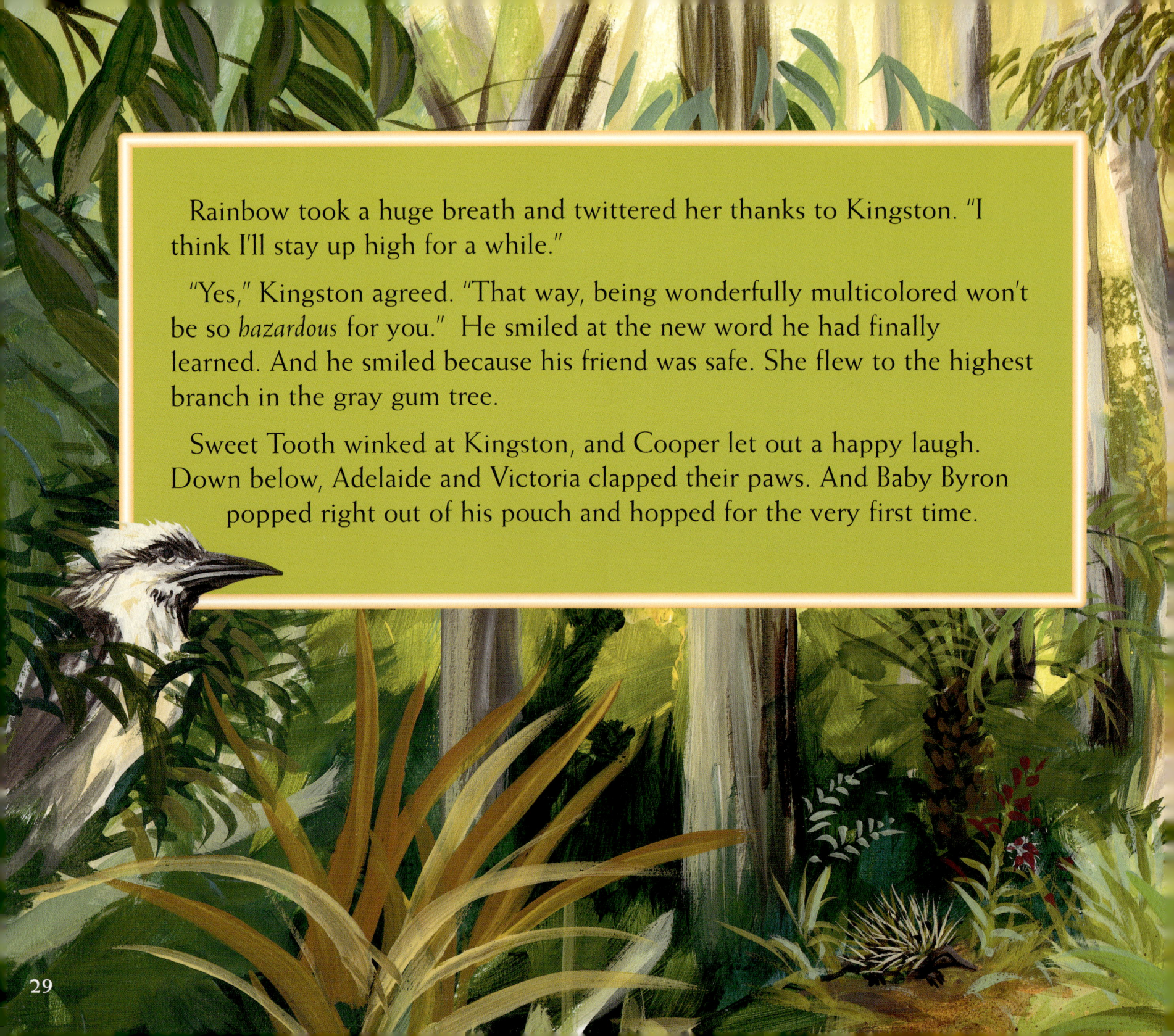

Rainbow took a huge breath and twittered her thanks to Kingston. "I think I'll stay up high for a while."

"Yes," Kingston agreed. "That way, being wonderfully multicolored won't be so *hazardous* for you." He smiled at the new word he had finally learned. And he smiled because his friend was safe. She flew to the highest branch in the gray gum tree.

Sweet Tooth winked at Kingston, and Cooper let out a happy laugh. Down below, Adelaide and Victoria clapped their paws. And Baby Byron popped right out of his pouch and hopped for the very first time.

The **rainbow lorikeet** is a type of parrot known for its brightly colored feathers of red, blue, yellow, orange, and green. Some hunters capture rainbow lorikeets to sell as pets. The rainbow lorikeet eats fruits and flowers by crushing them with its sharp, strong beak and licking the nectar and pollen with its bristly, sticky tongue.

weeping bottlebrush

waratah flower

More About the Australian Temperate Forest

Koalas, wallabies, and the other types of animals featured in this story really do live in the temperate forests of Eastern Australia. They don't talk to one another the way that Kingston, Victoria, Sweet Tooth, and the others do in *Kingston's Flowering Forest*, but they face the same dangers and behave in some similar ways. Read more about these fascinating animals...

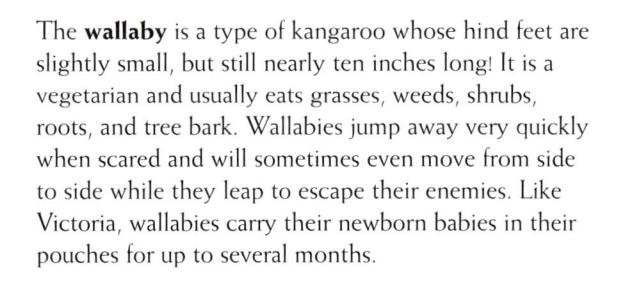

The **wallaby** is a type of kangaroo whose hind feet are slightly small, but still nearly ten inches long! It is a vegetarian and usually eats grasses, weeds, shrubs, roots, and tree bark. Wallabies jump away very quickly when scared and will sometimes even move from side to side while they leap to escape their enemies. Like Victoria, wallabies carry their newborn babies in their pouches for up to several months.

Echidnas, like Adelaide, are also called spiny anteaters. They are best-known for their sharp spines. When threatened or scared, an echidna rolls itself into a spiky ball and digs into the dirt with its long, strong claws. Echidnas' favorite foods are ants and termites. Along with the platypus, another Australian animal, the echidna is one of only two mammals on earth that lay eggs instead of giving birth to live young.

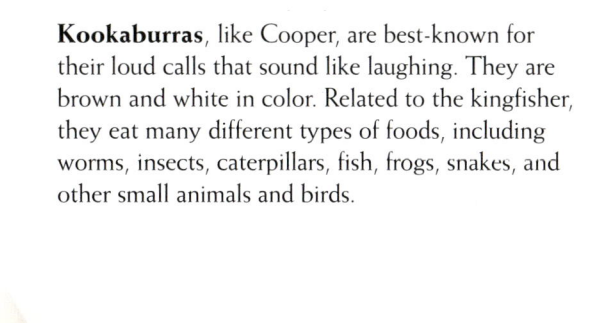

Kookaburras, like Cooper, are best-known for their loud calls that sound like laughing. They are brown and white in color. Related to the kingfisher, they eat many different types of foods, including worms, insects, caterpillars, fish, frogs, snakes, and other small animals and birds.

Flying foxes, like Sweet Tooth, are actually bats. While other types of bats eat insects, the flying fox eats fruits and flowers. Its favorite foods include mangoes, avacados, bananas, dates, figs and certain flowers. Some flowers that depend on the flying fox or other bats for pollination open at night when bats are active. The flying fox finds its food by sight and smell, unlike many other types of bats, which use a radar-type sense called echolocation.

Koalas, like Kingston, really do eat only the leaves of eucalyptus trees, also called gum trees. Since oils from the eucalyptus leaf are used to make cough and cold medicines, many koalas smell like cough drops! Koalas spend most of their time alone, high in trees, and sleep for up to eighteen hours a day. Their name comes from an Aboriginal word that means "drinks no water," because they get all the moisture they need from the leaves they eat.

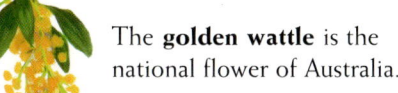

The **golden wattle** is the national flower of Australia.

Eucalyptus (gum tree)

Where in the World Is Australia?

Australia is an island in the **southern hemisphere**. Some of the **temperate forests** of **Australia** are called **eucalyptus forests**. Australia has more than **400 species** of eucalyptus trees.